A COMPLETE MANUAL
OF THE
EDISON PHONOGRAPH.

By GEORGE E. TEWKSBURY.

WITH INTRODUCTION
By THOMAS A. EDISON.

PREFACE.

THE preparation of this modest work has been undertaken for two reasons. First, there is no guide to the standard Edison Phonograph as now manufactured in its perfected form. Mr. Andem's excellent book describes in graceful phrase the machine of several years ago, before the new shaving device and other changes had been adopted, and before the day of the spring motor. The instrument has since been perfected, and the field of its usefulness broadened. Improvements in motors, batteries, speakers, the use of special glasses, and the advent of many new devices, daily suggest questions which had not then arisen. The art of record-making likewise has advanced, and more inquirers, amateur and professional, want to know about this fascinating employment.

The attempt, therefore, is made to give practical directions in plain language as to various operations which an every-day experience of nine years has suggested or invited, and which are the subject of constant inquiry, particularly from beginners, such as adjusting the tension of the twin-nut spring, setting the diaphragm, the building-up of speakers, how to shave, the kinds and use of horns, the way to make records, the adjustment of sapphires, and other kindred subjects.

Secondly, to save time, impossible as that may sound. For these pages have not been written in moments of idleness, but in the strife and stress of busy working-days, when sometimes the hours seemed too few for the daily task. To save time, then, by answering questions more faithfully than the hurry of a daily business mail would allow, when indeed many things could be touched upon not at all; to explain what obscure causes will produce simple results; to help where help is needed; and to make easier and more delightful that enjoyment which the great genius of Mr. Edison, and the work of his followers, has made possible.

CONTENTS.

ILLUSTRATIONS.

THE EDISON PHONOGRAPH.

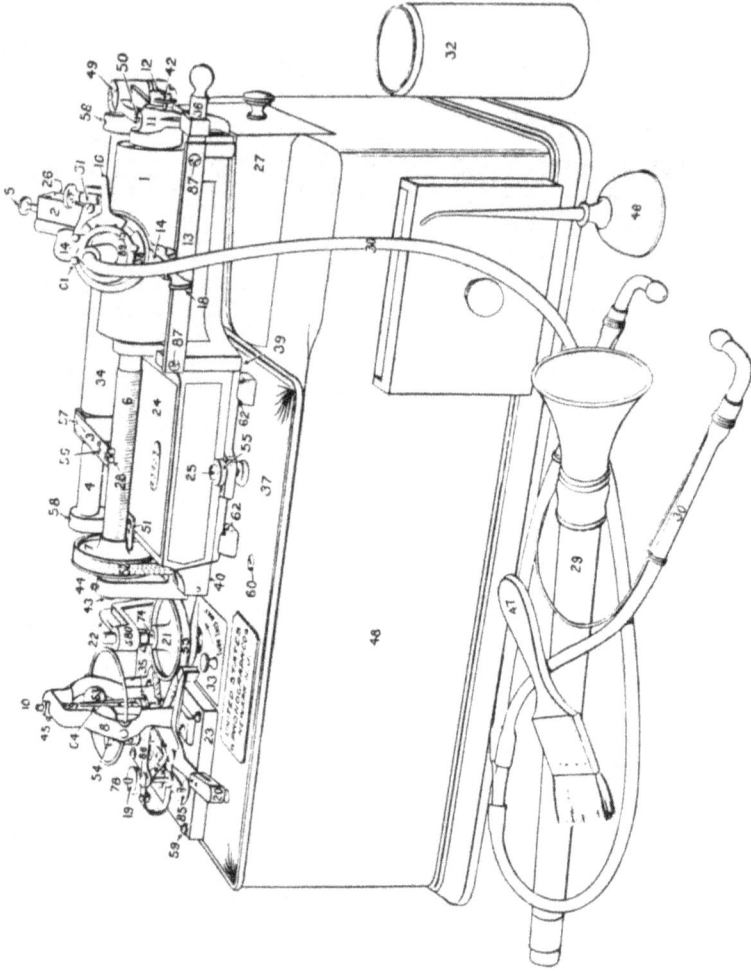

EDISON PHONOGRAPH — INDEX OF PARTS.

INTRODUCTION.

SOME years ago, I " wrote a piece " about the Phonograph, for the North American Review. Nine years ago, that was— further back than most men like to remember, and a long time in the history of an invention. Since then many things have happened, in science, in experiment, and in every phase of human activity. In looking over that article, I do not see that there is a great deal to change in it, even at this distance of time. Much that was then anticipated has come true, and much more than was then expected has happened. But the eye of prophecy always sees dimly. Merely to have foreseen a result or a general effect is enough, and it would have been the part of wisdom then, as it is now, to leave the details of fulfillment to the judgment of those taught by later experience, and the trend of the years.

Having arrived at an age when he believed men were little likely to change their minds on any important subject, Mr. Ruskin, reviewing an earlier work, said he would omit some parts but not attempt to mend. So in writing this Introduction for friend Tewksbury's book, I will refer only to those parts of the former article which seem to me of permanent interest.

I remember distinctly that I enumerated the reproduction of music for popular enjoyment, as among

the important uses that the Phonograph would
serve, and so widen the gentle influence of melody
and add to the general sum of pleasure.   This it has
certainly done, and is doing.   In addition, by bringing
within the reach of everyone accurate reproductions
of the best music, it exerts, almost unconsciously,
an elevating moral influence.   Dictation without the
aid of a stenographer, has been successfully achieved,
while a use that did not then occur to anyone,
namely, rapid transcription, has grown out of this
feature of the machine.   The teaching of elocution
and the study of languages were mentioned, and in
both of these fields I see increasing uses and larger
opportunities.   I know of no other such aid to
education.   The method and application are obvious.
Music-boxes and toys were mentioned; but with the
advance made in the art of record-making, the Phonograph is now so much better than any music-box, so
much more varied, truer to life and nature, and more
human, that it has left behind the older and more
mechanical device.

In anticipating that the Phonograph would be a
means of private entertainment, it did not occur to
me that public entertainments would become a
special feature of its use, and the instrument was
designed with no such end in mind.   Yet I find that
a large number of persons, not possessing an instrument themselves, receive satisfaction and pleasure
from its exhibition by others.

I am still attracted towards the Phonograph for the
same reasons that influenced me at the beginning.
In the material business world it is a time-saving

device of unlimited service, and in the sphere of music it more nearly satisfies the natural desire for song and melody than any other mechanical agent.

When my first "piece" was written the Phonograph was in its childhood. It seems now to have arrived at a vigorous maturity. In one sense it knows more than we know ourselves, for it retains the memory of many things which we forget, even though we have said them. It teaches us to be careful what we say, and I am sure makes men more brief, more business-like, and more straightforward. Sometimes I think it cultivates improved manners, and I feel sure that any means tending to perpetuate the charm of music must be a help and a solace to all.

For the future it has great possibilities, and the improvements that will continue to be made from time to time will in the end bring it to a perfection that will make the Phonograph an important feature of every household.

THOMAS A. EDISON.

TYPES OF PHONOGRAPH.

THE standard Edison Phonograph as now manufactured is made in two forms, known as the Electric Phonograph, or class M machine, and the Spring Motor Phonograph. Other types such as the treadle machine, the water motor machine, the C machine (built for use in England), were formerly put out, but now are abandoned or have become obsolete. Mr. Edison has lately brought out a cheaper type of machine called the Home Phonograph, but we shall concern ourselves in these pages with the standard instrument only. Different types described.

A Phonograph with motor specially wound for an electric light current of 110 volts, so that the machine might be operated from electric light wires, was made at one time, but the sale was limited, and we need only mention it here.

In considering the Edison Phonograph, it should be borne in mind that every machine, whether spring or electrical, consists of two distinct parts, called respectively the motor and the body. The motor is the driving or actuating mechanism, while the body or top is the speaking or Phonographic part. The motor is that part of the machine below the top plate, including of course the governor. The speaking part with its various mechanical devices, together called the body, rests Two distinct parts of the Phonograph.

upon the top plate of the motor, to which it is attached.

The electrical machine.

The Class M or Electric Phonograph consists of a perfectly adjusted electric motor, combined with a mechanism which accurately records and reproduces human speech, music, and other forms of articulate sound. The motor requires a steady cur-

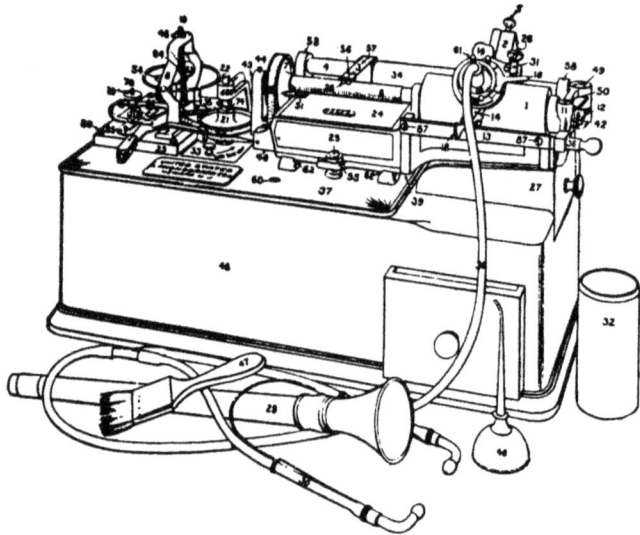

Secondary batteries preferred.

rent of 2½ volts and 2 amperes. Secondary or storage cells are generally preferred for this work. and when they can be readily recharged from a dynamo (as they must be) are undoubtedly the most satisfactory in ordinary service. But storage batteries being fragile require skillful handling, not merely in transportation but while recharging and in use. They are easily damaged, expensive to ship, heavy to carry, and therefore more likely to suffer from disarrangement and defects. Where no facilities are at hand for recharging, or where the conditions of use prohibit a bulky cell, a portable primary battery can be obtained that will do the work well. The advantages are instant availability, and lightness of weight. There is little choice in point of care required, although the portability of the latter may commend them to many. These matters will be more fully treated in a subsequent chapter, devoted entirely to the subject of batteries and motive power.

Advantages of primary battery.

For continuous service, or in the hands of a user familiar with electrical work, or of one who will give ordinary care to the apparatus, the class M Electric Phonograph is an ideal machine, than which no better or more beautiful instrument was ever made for any purpose.

The weight, lying largely in the field magnets, is 65 pounds. The instrument sits in an oak body box, as shown in the engraving, and is accompanied with a single hearing tube, speaking tube, oil can, and camel's-hair chip brush for cleaning the cylinders. Batteries are not a part of the Phonograph as manufactured, but are sold independently and as an extra.

Weight and accessories.

The Edison Spring Motor Phonograph is composed of the body or top of a standard M machine, in combination with a powerful noiseless clock-work movement that can be wound by hand, and will run through fourteen records with a single winding. The Phonograph top is attached to the spring motor in the same way as to the electric motor, and the results

Spring motor Phonograph described.

Cut of spring
motor,
Phonograph.

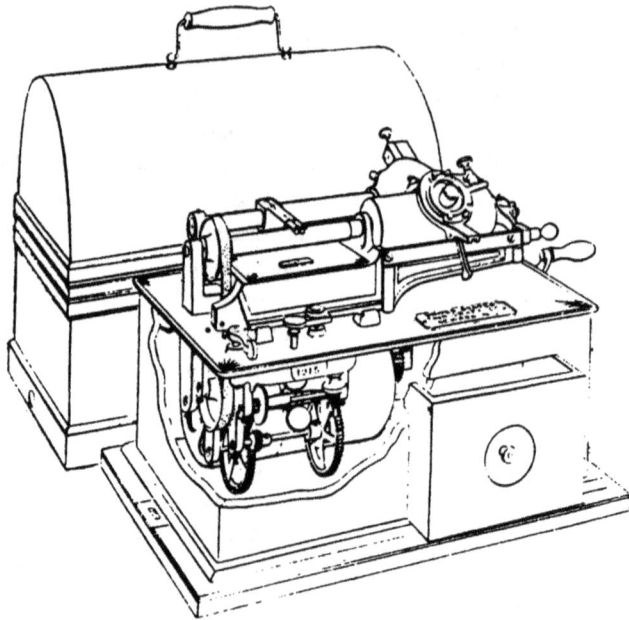

Description of
spring motor.

are no less satisfactory. The Edison spring motor consists of three powerful springs actuating a train of gears controlled by an ingenious governor, which silently revolve the shaft of the Phonograph, and drive the recording and reproducing arm of the machine. The Spring Phonograph is commended for portable service and home use, and particularly to those who fear the annoyance and troubles that sometimes proceed from the use or neglect of an electric battery. There is no difference between the electric and the spring machine in workmanship, as both come from the same skilled hands and are subjected to the same severe tests.

The Edison Spring Phonograph, with its body box and cover complete, weighs 42 pounds. It includes a speaking tube, hearing tube, oil can and camel's-hair brush. *Weight of spring machine.*

An English writer of note has said that Mr. Edison's Phonograph is built with the precision of a scientific instrument. It is probably true that no invention ever offered to the public has stood the test of use better. In the hands of inexperienced persons, harm seldom comes even to the most delicate parts, while the wear and tear, not to speak of neglect, to which the machine is subjected by professional record-makers, seem not to hurt it nor impair its efficiency. Every part of the instrument is built to an established standard, like a good watch. The user in India can send for a screw, a shaft, a center, a rod, a spring, a brush, and it will fit perfectly in its place when he gets it. *Accuracy of workmanship.*

With ordinary care the Phonograph will outlast the life of the man who operates it. It is not likely to get out of order, but if from any cause it should become disarranged the adjustments are so simple that they can be made by any person of intelligence. *Durability.*

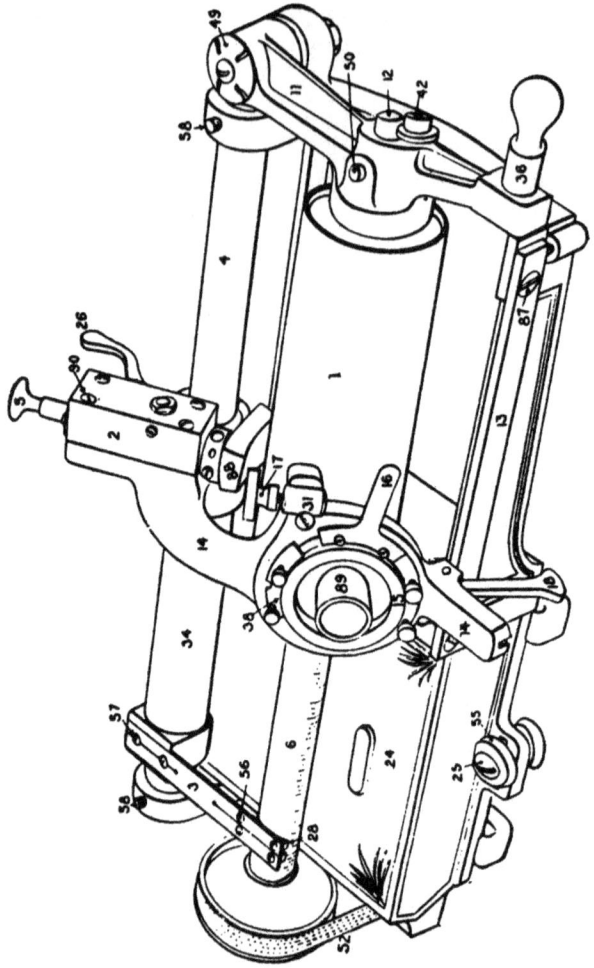

THE PHONOGRAPH BODY.

THE principal working parts of the Phono- Parts of the body. graph body are the main shaft (6) and mandrel (1), speaker arm (14), carrying speaker (15) and turning rest (2), back rod (4), back-rod sleeve (34), twin nuts (28) and spring (3), lift lever (18), swing arm (11), lock bolt (36), and swing-arm and main-shaft centers.

Every Edison Phonograph is equipped with a Equipment. standard speaker having sapphire points for recording and reproducing, and also with a turning rest or shaving-off device, with which the surface of cylinders may be planed, thus permitting the use of the same blank cylinder many times.

The speaker is the part of the Phonograph which The speaker. records the talk and does the talking. These functions are both performed by the same speaker, by adjusting it to one position when recording and to another when reproducing. The speaker (15) consists of a metal frame enclosing a sensitive glass diaphragm clamped at its outside edge between two soft rubber cushions. To this glass is connected a lever carrying two sapphires, one of which has a sharp edge that cuts the record and the other a polished How sound waves are recorded. ball for reproducing. The sound waves of speech or music vibrate the glass diaphragm, which in turn communicates the vibrations to the recording sapphire resting upon a cylinder of metallic soap, commonly

called a wax cylinder. This cylinder revolves
beneath the recording stylus, and indentations cor-
responding in character with the sound waves which
actuate the diaphragm are thus cut in the cylinder.
In reproducing, the sapphire ball of the speaker,
brought into operative position by the simple move-

*How sound is
reproduced.*

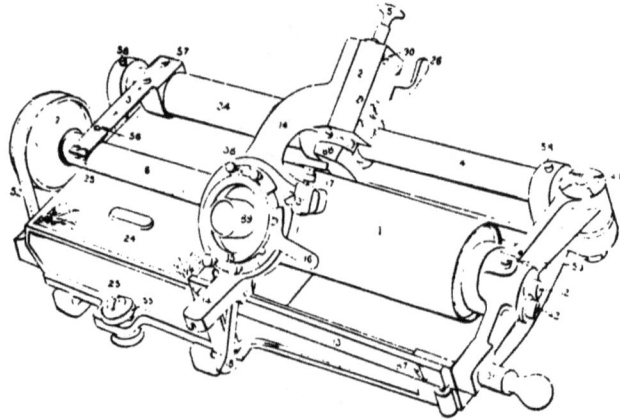

*Cut of the
Phonograph
body.*

ment of a lever, follows in the groove previously made
by the recording stylus, and vibrates the glass
diaphragm in the same manner as when the record
was recorded, causing the same vibrations and therefore the same sound.

*The shaving-off
device.*

The turning rest or shaving attachment (2) is fixed
to the speaker arm of the machine, and has an automatic adjustment. The principle is the same as the
tool of a turning lathe. The sapphire knife is held
in place against the surface of the cylinder, into which
it slightly projects when set, cutting off a thin shaving
as the wax cylinder revolves and passes under it.

Further information in regard to the practical use of the speaker and the turning rest in operation will be found under separate headings devoted to those important subjects.

The main shaft (6) is a fine screw 100 threads to the inch, assembled with a tapering brass mandrel (1) on which the wax cylinder is placed. Engaging with this is a pair of twin nuts (28), that feed the carriage. The proper tension of the twin-nut spring (3) is essential, because if the twin nuts bear too heavily on the feed screw they soon wear out, besides putting an unnecessary resistance on the shaft. In case of the spring motor machine it is even more important that the tension of this spring should be right. This tension is equivalent to two thicknesses of writing paper between the straight edge (13) of the Phonograph, and the lift lever (18) when dropped to its lowest point. To set the twin nuts and spring, first loosen the long-headed screw at the back of the speaker arm. When this screw has been released, the twin-nut spring and back-rod sleeve (34) will turn independently of the speaker arm. The twin nuts should now be set. There are slotted holes in the end of the spring for receiving the twin-nut screws. Place the twin nuts with the narrow shoulder toward the front of the machine. Then partially set up the four twin-nut screws which are to hold them to the springs, leaving them free to move in the slots. Lay two thicknesses of ordinary writing paper on the straight edge (13), drop lift lever (18), and hold down the speaker arm with the thumb of the left hand. With the little finger of the same hand,

The main or feed shaft.

Tension of twin-nut spring.

How to set twin nuts and spring.

gently press the loose twin nuts into the thread of the main shaft. Still holding the speaker arm and twin nuts down, run the machine a few revolutions, and the twin nuts will adjust themselves to the arc of the shaft. Stop the machine, still maintaining the equal pressure on the twin nuts and the speaker arm, and holding them thus, tighten the twin-nut screws and the big screw at the back. Withdraw the writing paper, and the tension will be right.

Adjustment of centers.

The main shaft is held in place by adjustable steel centers. The adjustment of these centers is very simple, but is often overlooked. The centers should be snug in their place, but not tight enough to bind. There should be no end-shake to the main shaft. If this shaft is loose enough to tumble on its centers the sound will be communicated to the cylinder when a record is made. If records are attempted to be reproduced on a loose mandrel, the eccentric motion of the cylinder will cause an uneven reproduction. On the other hand, if the centers are too tight they act as a brake on the shaft, retarding its motion and making the reproduction unsteady. If the main-

Loose main shaft and tight centers.

shaft centers are slightly oiled they will set snugly into place without affecting the free motion of the shaft, which can be spun by the hand when the main belt is thrown off the main-shaft pulley (7). A loose main shaft is among the commonest sources of trouble in recording, reproducing and shaving, when it would not be if occasionally looked after. Records, whether of speech or music, being made at a given number of revolutions per minute must be reproduced at the same speed. If at a higher speed the

tone becomes shrill and sharp, and if at lower speed
the pitch drops.   Hence it is necessary that the main
shaft of the Phonograph which revolves the cylinder
should have a constant motion without friction.

The swing-arm center (12) is adjusted by first
loosening the set screw above it, then turning the
adjusting screw which works in the slot at the end of
the center, and, when proper adjustment has been
obtained, tightening the set screw on top.   The main-
shaft center at the left hand can be moved by loosening its set screw.   See that set screw is firmly set up
after setting center.   The main-shaft stop, shown in
the main drawing and numbered 51, should be so
placed as not to touch the inside of the pulley (7),
and yet close enough to keep the main shaft from
leaving its left-hand center when swing arm (11)
is open.

The main-shaft screw (6), and the back rod (4),
should be kept well oiled, and none but the best
quality of Phonograph or clock oil used.

Never attempt to slide the speaker arm from side
to side without either raising it or closing the lift
lever, as you are liable to damage the thread upon the
main shaft by scraping the twin nuts across it.   Do
not remove main shaft from machine unless absolutely
necessary, and then use great care in drawing it out to
avoid injuring the thread.

Adjustment of swing-arm center.

Oiling feed screw and back rod.

Removing main shaft

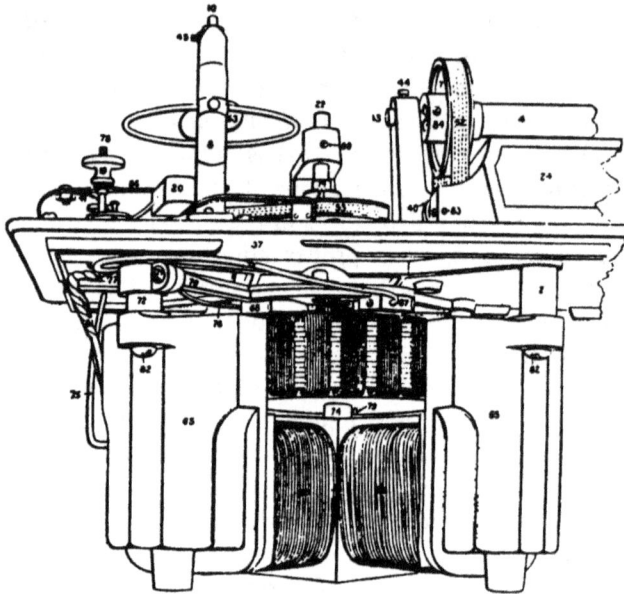

THE MOTOR AND GOVERNOR.

VERY little attention is required to the motor proper of the Phonograph, but as the subject of the motor includes the matter of brushes and the governor, some space must be devoted to this feature of the instrument.

Working parts of motor.

The two principal parts of the motor are the field magnets and armature. With the field magnets (65) we need not deal, as in practice no adjustments are required. The armature (66), which is the revolving "wheel" within the magnets, includes the commutator (69), which may be called its hub, in contact with which the motor brushes (67-68) rest. These brushes differ only in respect of the way in which they are attached to the connection wires. One of the wires runs to the binding post, and the other to and through the governor. The first is called the flat connection brush (68), the second the pin connection brush (67). Both are held to position on a rocker arm (70), which swings on the pillar (72) at the corner of the magnet. See illustration. The brushes at point of contact quarter the circle of the commutator. Sometimes the Phonograph is shipped with brushes drawn away from contact with the commutator, to prevent possible injury. If machine is thus received it is necessary only to press the rocker arm (70) inward until the brushes rest gently against the hub. The screw (71) clamps the rocker arm to the pillar. It may be loosened if necessary for readjusting. When loosening this screw it is important that the rocker arm be not allowed to drop from its original position, or there is danger of hitting the revolving armature, and thus perhaps breaking or disarranging the wires. In case it is ever necessary to re-solder one of these small wires, the soldering should be so nicely done as not to form a metal connection with any other wire. The only time the rocker arm need be moved is for cleaning or changing brushes.

The brushes must not press too tightly against the commutator, as crowding does not add to the value of the electrical contact, but creates friction, wears out brushes, and is sure to disturb their relative position to the commutator.

Oil fatal to electrical contact.

Cleanliness, however, is of the first importance. Dirt or oil act as a resistance, and retard the flow of the electric current. These parts, meaning the commutator and the copper brushes, must be perfectly dry and bright. Beginners are cautioned against oiling either of them. If sometimes oil creeps down from ˌthe oil cup at the top of the armature shaft, remove it without delay. When the brushes or the commutator become dirty, benzine is used for cleaning. A small piece of cheese cloth or other light fabric saturated with benzine and held against the commutator with the finger, through the opening in the top plate, will remove the dirt. Turn the armature to the right with the hand, thus covering the entire surface. The brushes are cleaned by first loosening the screw (71), and swinging out the rocker arm to bring the brushes in view. Use a piece of cloth as above, or a soft toothbrush, drawing or brushing towards the end. Use care not to spread or displace the fine copper wires. If these become rough or uneven, a fine file will brighten and smooth them, in which case the copper filings should not be permitted to remain in the brushes. In putting the brushes back in position against the commutator, which of course should be cleaned at the same time, observe the caution mentioned in a preceding paragraph, and do not crowd them out of place against the segments.

Way to clean brushes.

Keep the armature-shaft center (22) snug in its place to prevent side motion. The jar of transportation will once in a while loosen the armature (66) on its shaft, permitting it to rest on the field magnets (65). If this should occur the armature can be put back in normal position by means of the screws which hold it to the shaft. The top plate is removed by taking out the four top-plate screws (60), which set

Armature shaft should be snug.

How to remove top plate.

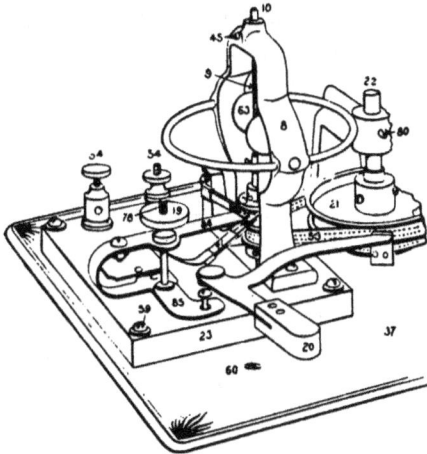

in the four pillars of the magnet, and loosening the armature pulley.

The Phonograph governor is very sensitive and very perfect. With ordinary care, it will not get out of order. The governor shaft (9) is held in place by the center (10), which is itself held by the set screw (45). The shaft should always be just snug, with no up-and-down shake—that is all. There are two electric brushes operating

The governor.

with the governor, and they are called the governor top, or speed, brush (41), and governor side, or
safety, brush. The use of the side brush need
not be explained here, as it requires neither adjustment nor attention. The top brush (41) controls
the governor, and therefore the speed of the
machine. It rests against the under side of the
copper disc (35), called the governor contact,
and must never touch the hub of the governor contact. If it touches there the motor will
run wild. To increase the speed of the Phonograph, turn the speed-adjusting nut (19) to the
left; to decrease it turn to the right, until the pitch
desired is obtained.

The governor brushes, particularly the top brush,
should be kept perfectly clean and bright. It is of
the greatest importance that the governor-contact
disc (35) should also be clean, so as to permit the free
passage of the current. Owing to its position on the
shaft it is sometimes difficult to keep oil away from
it. Oil should therefore be applied sparingly to
the top center of the shaft, and on the shaft where
the sleeve works, if needed. To clean governor
brushes and governor contact, remove top center
(10) by loosening set screw (45), and lift shaft out
of its place.

If the governor belt (53) is too loose, it will slip on
the governor pulley, causing imperfect regulation.
Lumps of wax or dirt on the governor pulley or
governor belt will produce the same effect. A
method of tightening this belt is provided, as
explained on page 37.

The use of the start-and-stop switch is too obvious to need explanation.

The ends of the battery cord are connected to the binding posts (54) at the back of the governor block. It makes no difference which of the battery wires runs to either post.

Attaching the battery cord.

OPERATING THE MACHINE.

Practical
directions.

BEFORE proceeding to the discussion of special subjects, it is deemed desirable, in behalf of those whose acquaintance with the Phonograph is not complete, and in some cases perhaps entirely recent, to publish directions for operating the machine. They will be of value to the beginner, and possibly of use to older enthusiasts—for all of us are enthusiasts, varying in degree according to the length of the courtship. The novelist told his readers at the outstart that those who were not interested should turn to certain pages, naming them, where he had prepared a dish that would tickle their jaded palates. So now, if thou dost not care for this simple diet, look further and mayhap thou wilt find the savory morsel.

Three functions
of the machine.

The happy owner of a Phonograph wants to know a great many things at first, but chief among them are these three : how to record, how to reproduce, how to shave. Here are the rules, referring to the numbered diagrams.

How to record.

To record. Machine at rest, with lift lever (18) pressed up. Throw down lock bolt (36), and open swing arm (11) wide. Slip wax cylinder (32), beveled end foremost, upon the tapering brass mandrel (1), and press it firmly, but not too forcibly, into place. Close swing arm and relock it. Raise speaker arm (14)—an inch is sufficient—from straight edge (13), upon which it rests in front, and slide it to the left until directly over the beveled end of the cylinder, or the point at which you wish the record to commence.

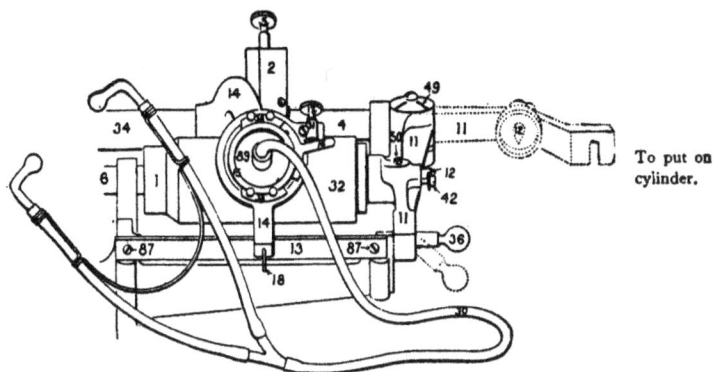

To put on cylinder.

Again lower it to straight edge, and draw cup lever (16) down as far as it will go. Start the machine by pushing brake handle (20) to the left until it strikes against the pin.

The machine being now in motion, place speaking tube (29) or horn upon the tube plate, lower lift lever (18) as far as possible, and the moment of recording

Recording position.

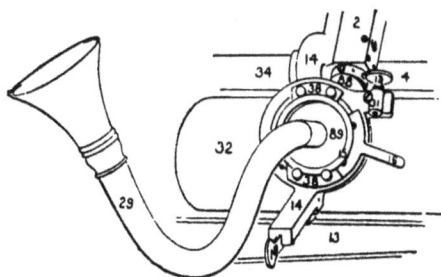

has begun. A fine white shaving will appear on the surface of the cylinder where it has been passed over by the recording stylus.

When the record is complete raise lift lever (18) to the highest point, thereby disengaging the arm and at the same time removing the cutting sapphire from the wax. Throw back speaker arm (14) as far as possible. Remove the shavings or "chips" from the newly made record by holding camel's-hair chip brush lightly against it, and passing it from left to right while the cylinder is revolving. Then shut off the electric current by throwing switch (20) to the right.

How to reproduce.   To reproduce. The machine is in motion, started by moving switch (20) to the left as far as it will go. Lift lever (18) up. Push cup lever (16) up against the point of adjusting screw (17). Now place hearing tube or horn on the tube plate, and lower speaker arm opposite the point where record begins, by dropping

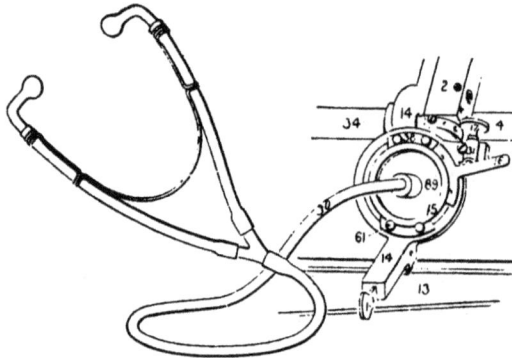

Adjustment to track of record.   lift lever (18). Although the reproducer ball usually adjusts itself to the track or groove made by the stylus, it sometimes occurs that clear reproduction is not at first obtained. To obviate this, unscrew adjusting screw (17) until its point disappears in the sleeve (31), and, while listening with the hearing tube, press cup lever (16) upward with the thumb of the right hand, and with the first and second fingers of the same hand slowly turn adjusting screw (17) down until you hear the record distinctly. This adjustment brings the reproducing ball into the groove of the record.

The speed of the Phonograph in revolutions of the main shaft per minute is regulated by governor adjustment screw (19). To increase speed unscrew this nut, and to decrease it screw the nut down. Observe this carefully when reproducing music, as a different speed from that at which the music was recorded will reproduce it at an entirely different pitch. The standard speed at which musical records are taken is about 125 revolutions per minute, and this speed is the most useful and reliable for all purposes. The governor speed (top) brush (41) on speed-adjusting spring should rest against the under side of the contact disc (35) only. It must never touch the hub of the disc, or the machine will " run wild." *Speed of the machine.* *Speed for reproduction.*

The shaving comes first, but we put it at the end, because the turning rest (2) is not, strictly speaking, an active element in the workings of the Phonograph, but merely an appliance for greatly increasing the capacity of the cyclinder. Yet everybody wants to shave, or thinks he does, which is much the same thing, and we will give a brief explanation of the device, and the reasons of its use, in these details of instruction. Elsewhere we shall enter into the more scientific field of shaving for professional record-making, and afford the reader, we trust, a more interesting account of the devices and expedients resorted to to obtain that " surface " without which a record can have little musical value and no wearing quality, *Shaving cylinders.*

and the secret of which has baffled all of them except a few.

The machine is now at rest. Wax cylinder firmly set upon mandrel; hearing tube removed; cup lever (16) set as for reproducing (up). Fasten back the speaker weight by passing rubber band around lower

To shave cylinder for another record.

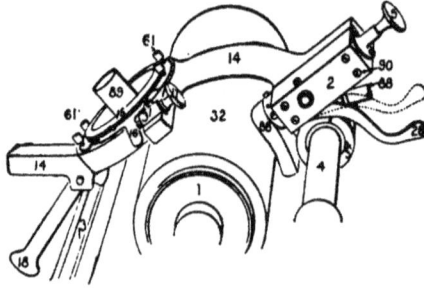

end of the weight and over the speaker arm about where the figures 89 are shown in cut. Remove speaker, if preferred. Lower speaker arm and lever (18) about over center of cylinder. Hold end of the speaker arm (14) down firmly with the thumb and forefinger of the left hand, while with the same fingers of the right *gently* press the button (5) of knife lever downward and toward the machine. Then press down knife lever (26) as far as it will go, close lift lever (18) up into its slot in the speaker arm, slide the arm to the extreme left, again lower lever (18), and start the machine.

Entire length must be shaved. The knife should always be allowed to pass over the entire length of the surface of the cylinder, otherwise there will remain a portion of the wax, which is thicker than the rest, and if a new adjustment of the knife be made to the right of the end of former cut, it will not touch the surface to the left of it. If adjusted to the left, on reaching that part which was before unturned, the knife will take too deep a chip and will tear, instead of cutting, the wax. After practice, the eye and ear of the operator will become accustomed to the sound and appearance of a proper cut, and readily detect anything wrong. The thinnest possible shaving will leave the smoothest surface and waste the least wax. Shave several times if necessary, in preference to a single deep cut. New blank cylinders require trueing, as they are likely to be eccentric and do not have prepared surfaces. In trueing these, set the shaving knife on the highest part, if any, of the blank. When once trued, they always remain perfectly cylindrical. Never attempt to set the knife while the machine is in motion. When the turning-off of a cylinder is completed, always see that knife lever (26) is thrown up and back as far as it will go.

New cylinders must be trued.

SPECIAL DIRECTIONS.

Tension of belts.

AS records must be reproduced at an even speed, it will be seen that the tension of the belts plays an important part in the operation of the machine. New belts are stiff, and likely to stretch at first, and old belts become oily and slip on the pulleys. If either of these conditions is present the record will waver or sound false. Attention is therefore directed to the proper care of these simple parts. When once properly adjusted to place, little further care is needed.

Free idlers.

The main or drive belt running from the armature pulley (21) to the pulley (7) of the main shaft, passes under the idler pulleys (40). These idlers should turn freely on their spindles, and not be allowed to accumulate particles of wax or other foreign substance, or the regular motion of the main shaft will be interrupted and the machine will "run false." The tension of the main belt is always to be observed, especially when belt is new. The belt must be tight enough not to slip over the main-shaft pulley when the pulley is held firmly in the hand and the machine is in motion. To tighten this belt, a belt-tighten-

How to tighten main belt.

ing screw (39) is provided, by turning which, after slightly loosening the body-holding screws (25), the belt can be drawn to proper tension. The body-holding screws should be afterwards set tightly in place again. The governor belt running from the armature shaft pulley (21) to the governor of the machine, if not kept at proper tension, will produce the same trouble that a loose main belt does. To tighten this belt, loosen the screws at the four corners of the governor block (23), and move the governor block to the left until proper tension has been obtained, then reset the screws. To tighten governor belt.

Belts should be kept free from oil, and if from careless oiling or neglect they become saturated with oil and slip on the pulleys, new ones must be obtained. To remove the governor belt, loosen set screw which holds the governor-shaft center (10), and lift governor shaft (9) out of its place. To change governor belt.

When we speak about tightening the belts, it must not be understood that the belts are to be drawn up rigidly, for one excess is as bad as the other. Remember the moderation counselled by the preacher when he prayed for rain. Belts not too tight.

A general caution to keep the entire machine perfectly clean and free from dust probably will not be observed. But it is essential. It is an essential to perfect work with any piece of machinery, and the Phonograph is no exception. The rules for oiling are simple. Phonograph oil should be applied *sparingly* every two or three days to the following parts. (*a*) The oil hole, marked on cap, near the back of the governor block. (*b*) Base of governor shaft. (*c*) Small hole in top of governor frame. A very little only, best put on with the end of a toothpick or a broom-straw. Note how small the receiving cup is at the top of the governor shaft. (*d*) Top-center bearing of armature shaft. (*c*) Bearings of idler pulleys under Cleanliness necessary. How to oil.

which the main belt turns. (*f*) End bearings of main shaft. (*g*) Thread of main shaft. (*h*) Back rod. (*i*) A little on the straight edge.

Too much oil means bad government.

Keep oil off the belts, and off the commutator and governor contact, or your motor will govern badly. Oil only where directed. In smearing it upon any other part you simply set a trap for dust and trouble.

The foregoing applies particularly to the Electric Phonograph. Directions for oiling the Edison Spring Motor are printed in the chapter devoted to the spring motor.

Copper brushes must be clean.

Note particularly the instruction under other heads regarding the care of the copper brushes, both motor and governor, which must be kept clean and bright. This caution is several times repeated, but never often enough.

Turn armature one way only.

It would not seem necessary to caution users against trying to run the armature backward, yet we have known of many cases where this was done out of curiosity, and the curiosity duly satisfied by the disarrangement of the wires in the motor brushes, the only remedy for which was the purchase of new ones.

Loudness governed by records and speaker only.

If trouble is experienced in not getting loud reproduction, the beginner's first thought is that the machine is not right. Numberless letters are written by exhibitors and others claiming to have especially loud Phonographs, referring to the machine itself. The actuating mechanism, provided it is new and in good order, has little to do with the fact complained about or boasted of. It is the records and the speaker that make a Phonograph loud or weak, clear or dim, the rest of the machine being merely secondary in that it provides power and means only. The speaker, then, should be looked to, provided inferior records are not being used. If the records and speaker are good, satisfactory results must surely follow. Do not tamper with the motor or the body to make your machine "loud." Look elsewhere, as explained here and in other pages of this Guide.

Confusion often arises in the mind of the novice from advertising claims by enterprising showmen. Let it be remembered that all standard Phonographs are alike, except as to motive power. There are no "grand concert Phonographs," except as users make them so; no "special trumpet Phonographs," except in the same sense; and above all no Phonographs "specially manufactured" for any purpose. *No special Phonographs.*

The adjustment of the wax cylinders is sometimes improved by removing them from the mandrel and replacing them in different position. If mandrel is not clean, or if inside of cylinder has any foreign matter on it, the cylinder will not revolve perfectly. *Resetting cylinders on mandrel.*

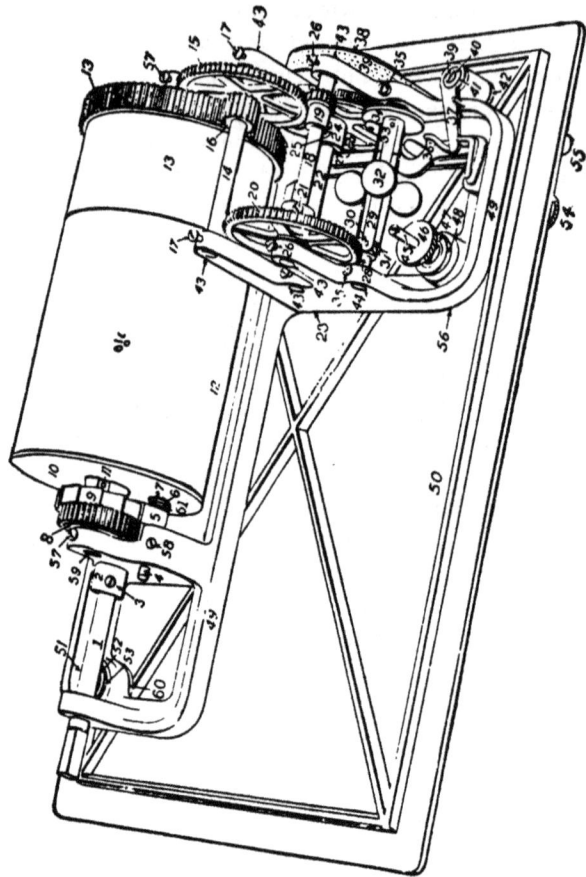

EDISON SPRING MOTOR — INDEX OF PARTS.

1 . . . Winding Shaft, assembled with Pinion.
2 . . . Winding-shaft Collar.
3 . . . Winding-shaft-collar Set Screw.
4 . . . Pawl Stud.
5 . . . Pawl.
6 . . . Pawl-stud Washer.
7 . . . Pawl-stud-spring Screw.
8 . . . Winding Gear ⎫ Assembled with
9 . . . Ratchet Wheel ⎭ Winding Sleeve, 11.
10 . . . Barrel Plate.
11 . . . Winding Sleeve, assembled with Nos. 8 and 9.
12 . . . Main Barrel.
13 . . . Driving Barrel and Gear.
14 . . . First Shaft ⎫
15 . . . First Gear ⎬ Assembled.
16 . . . First Pinion ⎭
17 . . . First-bushing Set Screw.
18 . . . Second Shaft.
19 . . . Second Pinion.
20 . . . Second Gear.
21 . . . Second-gear Set Screw.
22 . . . Third Shaft.
23 . . . Third Pinion.
24 . . . Third Gear.
25 . . . Second-pinion Set Screw.
26 . . . Second-bushing Set Screw.
27 . . . Third-gear Set Screw.
28 . . . Governor Shaft, assembled with Pinion.
29 . . . Governor Springs.
30 . . . Governor-spring Screw and Washer.
31 . . . Governor Collar and Set Screw.

32 . . . Governor Ball.
33 . . . Governor Sleeve ⎫ Assembled.
34 . . . Governor Disc ⎭
35 . . . Governor-bushing Set Screw.
36 . . . Governor Rocker, assembled with Leather Shoe.
37 . . . Switch Spring.
38 . . . Drive Pulley.
39 . . . Drive Belt.
39A . . . Switch-stud Set Screw.
40 . . . Switch-spring Washer.
41 . . . Switch Stud.
42 . . . Switch-stud Lower Washer.
43 . . . First, Second and Third Bushings.
44 . . . Governor Bushing.
45 . . . Belt-tightening Screw.
46 . . . Belt-tightening Thumb Nut.
47 . . . Belt-tightening Washer (metal).
48 . . . Belt-tightening Washer (rubber).
49 . . . Motor Frame.
50 . . . Top Plate.
51 . . . Motor-frame Screw.
52 . . . Motor-frame-screw Washer (metal).
53 . . . Motor-frame-screw Washer (rubber).
54 . . . Speed-adjusting Screw.
55 . . . Start-and-stop Switch.
56 . . . Belt-tightening Spring.
57 . . . Barrel-shaft Set Screw.
58 . . . Pawl-stud Set Screw.
59 . . . Barrel Shaft.
60 . . . Cushion Washer.
61 . . . Pawl-stud Spring.

THE SPRING MOTOR.

THE body or top of the Edison Spring Phonograph is the same as the top of the Electric Phonograph. We shall concern ourselves with the motive part only.

Spring motor parts described. The accompanying outline cut of the motor, with its index of parts, clearly mark and name all the visible mechanism. Within the barrels (12 and 13) are three powerful main springs, and arbors on which they wind. The arbors turn on the barrel shaft (59). The inside ends of the springs are attached to the arbors, and the outside ends to the barrels. The springs can be wound while the Phonograph is reproducing, without varying the speed of reproduction. By an ingenious device they wind and unwind one after the other, the power being conveyed to the Phonograph through a train of gears. The springs are held in tension by a noiseless pawl (5), so constructed as to be operative at either end, and a ratchet wheel (9).

Variety of motors. A variety of spring motors, generally bad, has been made for the Phonograph. The common fault is imperfect regulation, the one essential feature of Phonograph reproduction. The government of the Edison Motor is accomplished by a governor with perfect metal pinion, and a frictional disc, the movement of which is limited and held by a speed-adjusting screw (54), and an intermediate rocker, which also acts through the switch (55) to start and stop the machine.

The motor is mounted on a frame so connected Adjustable with the top plate (50) as to afford a hinge-like action frame. to the whole mechanism, which permits the belt to be tightened by turning a screw (46). The gear shafts all run in removable hardened steel bushings (43). The motor winds and runs noiselessly, without vibration. It is built to reproduce fourteen records when

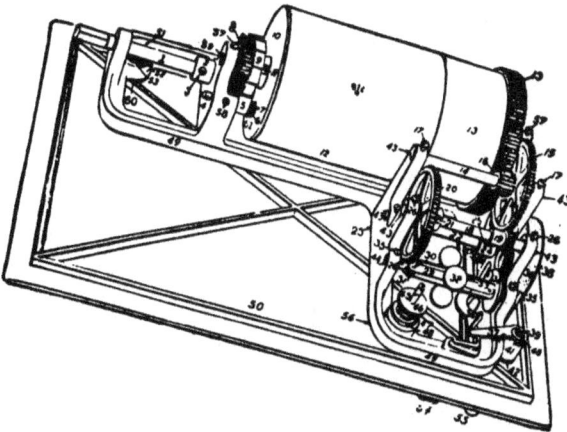

wound up. It will shave from 5 to 7 cylinders, although Capacity of when shaving the work is more quickly done by motor. keeping the springs at ample tension. In the spring motor, as in all other machinery, friction is the foe to power. See to it, therefore, that all parts of the motor are kept free from unnecessary resistance. Set screws in metal, whether on the electrical or the spring machine, or any other piece of machinery, are liable to work loose from the jolt of travel. When screws become loose, parts sometimes become disarranged.

Gears must be free.
The train of gears must always be free. The bushings, while themselves to be held firmly by their set screws, must not crowd against the shoulder of the pivots, or they will act as a brake on the machine, the same as a tight center. The shafts must turn freely. Particularly is this true of the delicate governor shaft, which is a most vital point, and responsive to the slightest load.

Tension of belt.
The belt must not be too tight, but loose enough to slip when main-shaft pulley of the Phonograph is held stationary with the motor running. If too tight there is a waste of power. It is hardly necessary to remind anyone that the drive pulley of the spring motor should not touch the frame, and should align perfectly with the main-shaft pulley of the Phonograph body. The belt must not run on the flange of the wheels. The tension of the belt can be regulated, as it stretches from use, by turning the thumb nut (46). Do not turn this when the machine is in motion, as there is danger of hitting the moving governor and disarranging the governor springs.

Cleaning the gears.
The teeth of the gearing should be kept free from dirt, for which purpose benzine and a toothbrush again demonstrate their value. Afterwards, apply a drop of oil. A fruitful source of trouble arises from the twin-nut spring being set at too great a tension on the feed screw of the Phonograph, practically checking the action of the motor. Full instructions as to the proper adjustment of this spring, have

already been given, and they should be followed. The Phonograph main shaft should not bind between its centers. Care should be observed to keep rubber cushions and washers in good condition. These serve to reduce vibration.

The speed of the machine is regulated by the screw (54). To increase speed unscrew the thumb nut, and to decrease it screw the nut down. To start machine, throw switch (55) to the left; to stop it, throw switch to the right. In shaving cylinders, best results are obtained by running machine at highest possible speed, unscrewing the nut (54). *Regulation of speed.*

It will sometimes happen that the coils of the springs, slipping on each other, make a jumping sound called " chugging." In most cases this will disappear as soon as the springs have worn into place. It has been found that oil gums the springs and increases the trouble. The springs, therefore, should be lubricated if possible with a dry lubricant, of which powdered graphite is the best known. Before applying the graphite wash the springs clean by standing the motor on end, with open end of the barrel uppermost, and pouring benzine through the coils as they unwind. When springs are dry wind them up, set motor on end as before, and while unwinding pour the powdered graphite in small quantity down through the springs, being careful to distribute it thoroughly. Only one or two applications are necessary. *Slipping of springs.*

Keep motor well cleaned and oiled, and free from dust. The following instructions in regard to oiling have been published, and we reproduce them here : *Oiling.*

Apply oil sparingly, but often, to the following parts, never over a drop at a time in any one place, except where otherwise specified.

The two bearings of the winding shaft.

The teeth of the ratchet wheel on which pawl works.

The inside surface of the friction disc of governor, where the leather touches it. This is important, and must never be neglected.

The cupped centers at the end of every gear shaft.

The governor shaft, where the governor-disc sleeve moves on the shaft, if dry.

Three places will be found for oiling the barrels, and to these places several drops may be applied. (*a*) The main hole is plainly marked on the large barrel. The other two are in the hubs. (*b*) There is a square opening in the hub of the loose barrel plate at the right-hand end. By turning the winding shaft the oil hole can be seen through this opening. (*c*) The oil hole in the left-hand hub, if not in view, can be found by allowing the machine to run part of one revolution.

Use best Phonograph oil only.

Caution.

When lifting motor out for oiling, never set it down on the front gears, but let it rest on the back edge of top plate and on the driving gear (13).

Springs to be kept partially wound up.

When transporting Phonograph or motor, wind the springs up several turns to hold them securely to their fastenings, thus insuring them against the effects of jar.

The Edison Spring Phonograph rests in its own box, which with cover forms a complete carrying case for the entire machine. By saving twenty pounds in weight of apparatus alone, beside weight and bother of batteries, it is recommended for portable service, while its perfect operation fully justifies the place it has won in public esteem.

BATTERIES.

BEFORE the invention of the spring motor, batteries were the only practical source of power for the Phonograph. The water motor was not a success on account of the unreliability of the water pressure, the plumbing required, and the fixed character of the installation. The treadle machine met with a fair degree of approval, but owing to improvements made in the storage battery about the same time, the electric machine soon replaced it. There was also a form of electric Phonograph with motor wound for a 110-volt direct current to be taken from the electric wires. This machine had some advantages, particularly at a time when power from other sources had not been developed. Various forms of motive power.

As originally made, the Phonograph was supplied with a primary battery of the Grenet type. The objection to this was its short life, its uncleanliness, and the care required to maintain it. The experiments in storing electricity in secondary cells were progressing rapidly, and as soon they had reached a commercial stage which made it possible to transport the cells with fair security, the storage battery at once supplanted the bi-chromate type, and the Phonograph was put out without a battery. As now sold, it does not include a battery. Grenet battery first used. Advent of the storage battery.

Progress has also been made since the earlier days

in adapting primary batteries to the uses of light motive power, and in this category the Phonograph is

Primary or chemical batteries. included. There are several batteries of the primary
type that can be used with satisfaction for the
Phonograph. The electrical requirements of such a
battery are practical constancy of electro-motive force
during the discharge of the cell. Portable service
imposes limitations as to weight, size, etc., and therefore being built small the portable primary batteries
do not last as long on a single charge as cells of
larger type. Where the machine is not intended to

Portable primary battery (left). Storage battery (right).

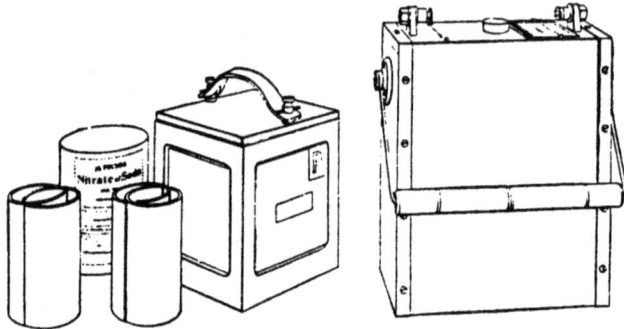

be moved about, and a permanent installation is
desired, larger batteries of the primary type may be

Advantages of primary type. used with satisfaction and good results. The advantage of a primary battery is that it can be recharged
by the owner, with renewals he can buy in the
market. In some cases this is desirable, and in
localities where there are no facilities for recharging
storage cells, which must be done from an electric-
light dynamo, it is essential. Using a primary battery,
and having on hand a supply of renewals for the same,

places the operator in control of the source of his battery power, and insures a current at all times.

Where there are conveniences for having batteries recharged at an electric-light plant, the secondary or storage battery is preferable. As its name implies, this battery stores the energy which it receives from the dynamo, and holds it until required. Various sizes within practicable limits are manufactured, each representing a certain number of Phonograph hours' service, but practical use confines itself to one or two sizes which can be easily handled and cheaply charged. Two storage batteries of practical size are preferable to one larger one, not merely on account of the difference in weight, but because they afford a more complete service, one charged battery always being ready to replace the other when it becomes exhausted. The energy stored in the battery does not go to waste when battery is not in use, but the battery will deteriorate rapidly if left uncharged. The acid attacks the lead plates and consumes them, if there is no electrical charge to protect the plates.

The component parts of the storage battery are two sets of lead frames, called the plates, immersed in a solution of dilute sulphuric acid, which must always cover the plates to the depth say of half-an-inch. As this solution evaporates, add water only, to maintain the proper depth of solution. Occasional renewal of the entire solution is advised about once in three months. If the battery has been leaking, or the acid spilled out, use new solution instead of water. To prepare the solution, use a good quality of commercial sulphuric acid of 66 degrees Beaumé,

Advantages of storage cells.

When energy will not waste.

Storage battery described.

The acid solution.

How to prepare solution.

in the proportion of five parts of water to one of acid, by volume not weight. Mix the acid and the water in a stone or glass vessel, and pour the acid slowly into the water; on no account pour the water into the acid, or it will cause an explosion. The acid solution should have a specific gravity of about 1190, or 23.2° Beaumé. Do not "spark" the cells nor hold a lighted candle, match, or other flame over the vent while charging.

The battery wires are attached to the posts projecting from the cell, which posts are the positive

Cut of primary battery for long service.

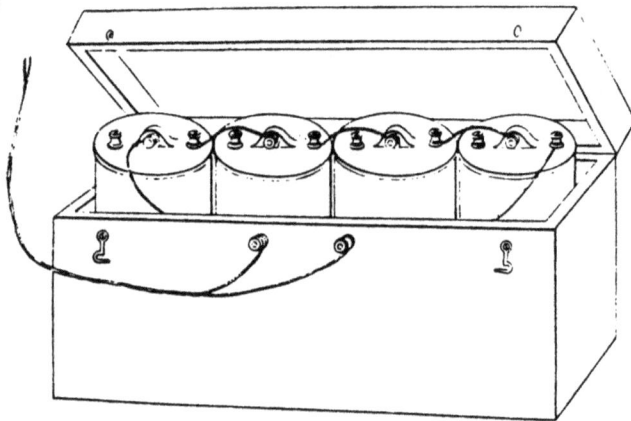

Wire connections.

and negative poles. It makes no difference to which of these posts either wire is connected. The posts should be kept bright and clean. If frequently washed with clear water and left to dry the corrosion will disappear. If permitted to remain unclean for any length of time, the posts will become so badly sulphated as to require the use of emery cloth to brighten them. When cleaning the posts, do not neglect the holes and screws. The binding posts being attached to the lead terminals of the battery, must be kept clean at that point of contact. It is a common occurrence for the acid so to increase the resistance there as to make the battery useless until it is cleaned. If neglected, the acid will entirely eat away the connection itself. Should this occur, new binding post must be soldered to the lug. The lead lugs may become eaten off where they join the plates below, in which case renewals are necessary. *Clean connections essential.*

Care should be taken when battery is not in use, that no metal substance is allowed to rest upon the binding posts, for if connected in this way what is called a short circuit is created, and the battery becomes immediately discharged at an excessive rate, creating heat to the melting point, with imminent danger of fire, and invariably buckling the plates and unfitting the cell for further use. When the discharge takes place through the battery cord and Phonograph in the usual way there is no danger, but a short circuit will be caused if the upper ends of the battery wire come in direct contact with each other. *Dangers of a short circuit.* The battery cord is heavily insulated, to prevent such contact, and if from any cause the insulation become worn off, the exposed parts should be wound at once with insulating tape, or otherwise protected. Keep the metal tips of the battery cord clean and bright, to insure perfect contact. *Insulation necessary.*

If the Phonograph will not start when the switch is thrown, the supposition is that the battery has run out. If the battery connections are found to be *When Phonograph will not start.*

tight, and the wires properly fastened, you can de-
To test if
battery is alive.
termine this by what is known as "sparking" the
battery. Hold the end of a short wire to one of the
posts of the battery, and sharply touch the other post
with the other end. A spark will flash out viciously
and bright if the charge is still there. If exhausted
or near exhaustion, the spark will be dull and
sluggish, or there will be none at all, and a freshly-
charged cell must be substituted.

Rough handling
fatal.
Jolting the battery, or handling it roughly, is likely
to break the inner hard rubber case on which the
plates rest, or do other damage. From this source
arise the dangers which attend upon the transportation of storage batteries by the express companies.
With ordinary care a good storage battery will last
two or three years, and sometimes longer, and the
battery will continue to yield its rated service. The
Maximum
efficiency not
obtained at first.
full efficiency of a storage battery is never obtained
from a new cell until it has been charged and discharged eight or ten times, when it will have reached
its maximum storage capacity. The first charge
should be continued for twice the length of time
usually required, and it is always advisable to put
into the cell about 10 per cent. more current than it
has delivered. Nearly all types of storage battery
have the nominal charging rate marked on the outside. It is from 10 to 20 amperes per hour, 10
amperes being the rate for the standard type now
commonly used.

Recharging.
All primary batteries contain printed directions for
recharging, and we need not explain the methods
here, but as storage batteries must be charged by

others, we condense the following instructions by one of the storage battery companies, for the benefit of electric light companies and others who have this work to do. How to charge storage batteries.

An alternating current *cannot* be used for recharging storage batteries.

To charge from an arc circuit, connect battery same as an arc lamp, positive to positive wire, using some dead cut-out.

Always switch battery on in circuit after starting, and off before shutting down.

Any number of cells can be charged at once by connecting them in series, negative to positive, and connecting end terminals same as one cell.

To charge from an incandescent circuit, it is necessary to use resistance in series with the battery, and an ordinary break-circuit switch. Connect positive terminal of battery to positive main, and negative to negative main, through rheostat and switch.

Here again, always switch battery on after starting, and off before shutting down.

To obtain proper charging current, connect sufficient resistance in series with your cells to register say 10 amperes on your ammeter. The resistance may consist of incandescent lamps in multiple, or coils of wire.

Never connect cells direct to dynamo mains.

To find resistance necessary, take voltage of system, subtract $2\frac{1}{2}$ volts for each cell charged at one time, divide remainder by 10 or 20 amperes (whichever is the proper charging current), and the result will be the resistance in ohms. Example: On 110-volt circuit, you desire to charge 2 cells at a time. Say the charging rate is 10 amperes. 110 minus 2 times $2\frac{1}{2}$, divided by 10, equals $10\frac{1}{2}$ ohms.

The length of time a cell should be left on circuit is easily determined by dividing its ampere-hour capacity by the charging rate. Before charging, open the cell and see that the solution covers the plates. If not, fill with clean soft water as previously explained, or acid solution if needed to replace leakage. Always open the vent before charging, and leave it open until charging is done.

NICKEL-IN-THE-SLOT PHONOGRAPHS.

Slot Phonograp.. described.

NICKEL-in-the-slot Phonographs are adaptations of the standard M Phonograph, with omissions and changes to meet the requirements of a coin-operated machine. The omissions are of parts that have no service to perform, such as body box, speaking tube, turning rest, etc., and the changes relate to the details of combination with another device. This device is required to enable the nickel dropped in the slot to start and stop the Phonograph, and to return it automatically to operative position after one reproduction. The first is generally called the nickel action, and the second the return mechanism. The Phonograph and automatic mechanism thus combined, and set in a cabinet, together form what is known as the nickel-in-the-slot Phonograph.

Principle of the device.

Many devices for automatic service were exploited some years ago when the idea was first conceived, but so far as the writer is aware only one of them has survived and been perfected, and only one is manufactured now. This mechanism has made the nickel-in-the-slot business practical. The principle is the closing of the electrical circuit by means of a nickel, not through the nickel or by its weight, but by the action of its diameter. The weight principle was tried and found wanting, being too delicate a method, as any light balancing action must be when subjected to rough usage and varying levels. The idea of a slot machine being to provide a single reproduction for the coin deposited, it necessarily must be equipped with a reliable device to stop the machine when through playing, and return it to operative position for the next coin. This is accomplished by a return screw, in combination with a feed lever, which replaces the lift lever of the Phonograph. After the Phonograph speaker has been raised from the surface of the record by the operation of a cam roller, the lever feeds the speaker arm of the Phonograph back to its proper place, tripping out the nickel as it passes, thus breaking the electrical circuit and stopping the machine. The return mechanism, as this part is called, is driven from the main shaft of the Phonograph, by a belt running on the hub of the main-shaft pulley.

Return mechanism.

Coin chute.

The coin is conducted to the nickel action below through an ingenious open chute with a steel bumper, which throws out lead, paper, tin, etc., and the receiving slot of the nickel action itself being open at the bottom discharges a smaller coin than the nickel for which it is set.

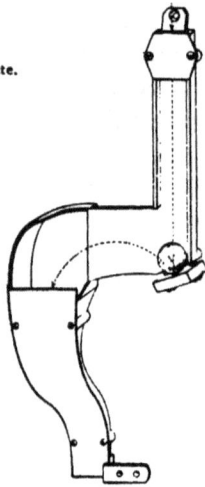

While nickel-in-the-slot Phonographs have proved profitable when distributed about in prominent locations, the success of the business has been achieved, and is now maintained, in what are called parlors, where a large number of machines are grouped together, and where greater pains can be exercised

Phonograph parlors.

in the selection and care of records, and in attention to details that will attract the public. The interest in the machine does not then depend on a single song or other musical selection, and the inducement to hear a number of selections is irresistible. Added to this are the attention which can be extended to customers and which customers like to receive, the more social and agreeable surroundings that such a place affords, and the opportunity given to women

Advantages of parlor system.

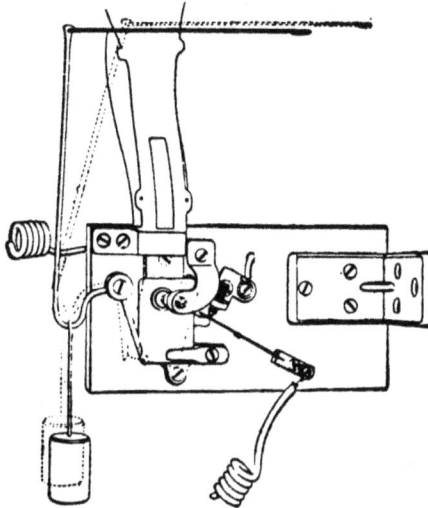

Nickel action.

and children to hear the Phonograph. The opportunity of making change adds to the receipts, and the enterprise is always under the eye of the manager.

Cleanliness about the machines and their accessories cannot be enforced too rigidly. The most successful parlors are those where in addition to the hearing of good musical records on the Phonographs, the cabi-

Attention to details.

nets are kept highly polished, the glass clean, the machines bright, the announcement cards fresh and interesting, the tubing white, etc., and in this particular it may be said that the Phonograph slot business does not differ from other enterprises that appeal to and depend upon the patronage of a scrupulous public whom it is not well to offend.

Requirements of slot device.

Any slot device to be effective must be thoroughly automatic, simple and reliable in action. Nothing is so injurious to the business as the failure of a machine to respond to the nickel. These points have been clearly set forth by Mr. Andem, the organizer and manager of several successful parlors. He enumerates the following as among the requisites of a good attachment. It should raise and lower the speaker arm gently at the beginning and end of the record, without jar or friction. Moreover, it should do this accurately. It should allow the Phonograph an

Cut of return mechanism.

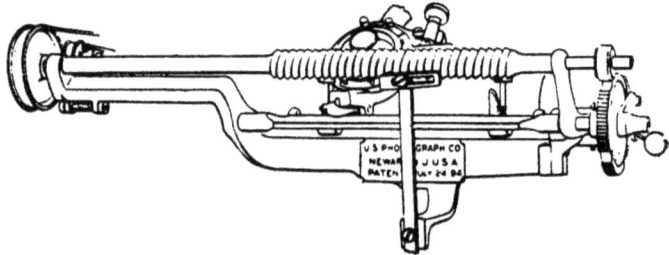

opportunity of a second or more for the motor to gain full speed and come under control of the governor, before reproduction begins. It should be so constructed as to raise the speaker arm and stop the reproduction before the electric current is cut off and the speed diminishes, to prevent the gradual dying-down of the sound.   It should be so attached to the Phonograph as not to put extra work upon the motor that will prevent the free starting of the machine.   It is also equally important that during the entire forward travel of the speaker arm during the period of reproduction, there should be no mechanical resistance from weights or springs, for besides the waste of power the drag will interfere with reproduction by binding the speaker arm and crowding the speaker out of track.   While reproducing, the Phonograph carriage should be as free as if there were no mechanism operating with it.   It should have electrical connections which are easily and surely made by the dropping of the coin, and strong enough to overcome dust and other ordinary resistance at the points of contact.   It should have a coin chute that cannot be clogged up, and a nickel action that cannot be reached and operated by wire or other devices that may be introduced through the receiving slot of the cabinet.   It should contain a tripping device that trips positively, and a discharge slot that acts without fail.   It should be adjustable to any length of record, and the entire device when once set in place should require no more attention than the rest of the machine.

Nickel-in-the-slot Phonographs are equipped with rubber, not brass mandrels, and automatic instead of standard speakers, the former to prevent slipping and breakage of records, and the latter to insure perfect "tracking" during reproduction.

Other requirements of slot device.

Receipt and discharge of coin.

Special equipment.

DICTATING AND TRANSCRIBING.

Different
methods
employed.

EVERY office man who uses a Phonograph for dictation has methods of doing his work which he believes more expeditious than anybody's else. But it has always been the author's opinion that no fixed rules could be laid down for the economy of this service, any more than for any other occupation of the human mind. The elder wisdom of many men and many minds would seem to apply. Yet there are certain suggestions that may be safely ventured.

Two machines
most economical.

The real economy lies in the use of two Phonographs for the office, so that dictation and transcribing shall be carried on independently of each other, and with absolute disregard of the work already in progress, or the engagements of the dictator and transcriber. The dictator should be free to use his Phonograph whenever he chooses, regardless of whether typewriting is going on or waiting to be done, and the typist or typewritist,

Reasons why.

or whatever new-born name they have for the clerk who operates a typewriting machine, should be equally free to use the Phonograph for his or her work without suffering delay when someone wants the machine for dictation. Hence two machines. This simple method secures the maximum of work, speed, economy and comfort, and a correspondingly greater efficiency, with absence of friction, and perhaps a sweetening of tempers all around.

We have not written to discourage the use of a single machine, but merely to suggest a better way.

The theoretical advantages of Mr. Edison's device for office use—a mechanical stenographer, perfectly accurate, ever ready, ever willing, ever cheerful, never tired, never sick, asking no salary and requiring no holidays—are innumerable, and the practical advantages many. It is doubtless a boon to have a faithful aid which makes no mistakes, because it cannot; contrives no lame constructions—at least none lamer than your own; has no objections, moral, physical or contingent, to working late at night, or even on Sunday if need be to save your neighbor's ox from the ditch. Still, one must not forget that the Phonograph is without brains, objective rather than subjective, as it were; hence possibly exacting finer mental activities on one's part, for which exaction, be it said, it repays tenfold and tenfold. *Advantages of a mechanical stenographer.*

But it is not difficult to dictate to the Phonograph, and it soon becomes a passion as well as an economy, creating a sort of official aristocracy, quite as the early use of typewriting machines did, when, as will be remembered by many readers, it was a business distinction to have "printed letters" instead of the other kind. A correspondent can dictate a better business letter to the Phonograph than in the old way. He is more precise, more accurate, more discreet. He thinks faster when there is no interruption, or, what is really the analysis of it, no fear of possible interruption; and thinking faster—not ham- *Easy to dictate to.* *Better results.*

Practical
advantages

pered by uncertainties and the hazard that does not,
because it cannot, respond to his moods—writes better
English, better sense, with more red blood in it. He
may change his speed, or he may leave his work and
come back to it, the machine is faithful—runs with
him, limps with him, stops for him, begins with
him. These all are practical advantages. Just
imagine for a moment, having a stenographer so
expert that from year's end to year's end he never
asks you to repeat a single word or a single sylla-
ble, and never halts upon a word. That is the
Phonograph.

Kinds of
machine for
office use.

When speaking of two instruments as the most
useful for office work, because of their greater time-
saving capacity and convenience, we had in mind and
would recommend a Spring Phonograph for dictation
or desk use, and an Electric Phonograph, if there is
no objection to it, for transcribing. The dictator
would use the machine that occupies the least space,
the one that would be ready under all circumstances
and therefore at critical moments, with no batteries or
battery wires about his desk, and requiring no
personal attention. The winding of the springs even
could be attended to, as we know of in one case, by
an office assistant at occasional idle moments if the
principal did not care to perform even this slight
labor.

The machine
for reproducing
only.

The transcriber's machine, equipped with an
automatic speaker for reproducing only, would be
more like a fixture, and with a battery service would
require no other use of the busy hands except the
ordinary manipulation. The shaving of cylinders,

to remove former dictations, and getting the blanks in Shaving off old dictation. order for another day's work, would be done on this machine during a few moments early in the day, or by an assistant; and on account of the greater speed which is possible by throwing off the belt of the Electric machine the work would be done speedily.

The process of dictating to the Phonograph is so Directions for dictating. simple as to need no explanation beyond that contained in the general directions for operating. We have said above that the machine will receive dictation perfectly at any speed. It will also stop altogether for you, and not lose its story. For in dictating letters a person has frequent occasion to stop abruptly, and in many cases loses the thread of his discourse. When you wish to stop abruptly in the midst of dictation, it is only necessary to close the lift lever (18), without stopping the machine. Should you have forgotten your last few phrases To stop temporarily, or correct. when you again resume, lower the lift lever, and, raising the cup lever (16), listen with the speaking tube. The action of the cup lever throws the reproducer into the track some five or six threads back of where the recorder stopped. After listening to the last few words, again bring the cup lever into position, and continue dictating (without stopping the machine). To avoid errors, the names of persons To avoid errors. and places should always be spelled out, unless the operator is familiar with them, the same as in ordinary stenography. The speed of the main shaft (6) for dictating should not be less than seventy or eighty revolutions per minute, at which speed it will take about four minutes to cover the entire surface.

Until the eye becomes practised, the number of revolutions of the main shaft can be determined by counting the revolutions of the set screw on pulley (7). Should the end of cylinder be reached before a letter is ended, it is only necessary to say "Continued," and finish upon another cylinder.

Not necessary to speak loud. It is unnecessary to speak loud when dictating. A distinct enunciation is the important thing. It is like talking through the telephone—the loud speech is not so well carried or understood. Hold the speaking tube within half an inch of the mouth, or rest one edge of it on the chin, speak naturally and and clearly, and articulate well. Talk to the machine precisely as you do to your stenographer, in the way of giving instructions. Say "Strike that out," or **Personal instructions.** "Change that to 25," or "Look up proper address," . or "Verify those figures from the books," etc. The transcriber always listens ahead, the same as he reads short-hand notes ahead before writing them out.

Transcribing. Transcribing is quite as simple. The mental process is identical with the present method from short-hand notes. Listening to the dictation, the transcriber hears a phrase, lifts the lever (18), when all is suddenly silent, writes it out, drops the lever again and goes on ; or if it is desired to hear the dictation repeated for correction or verification, pushes the carriage back to the starting point, drops the lever, and listens to the same thing over again. The hearing tube is not removed when transcribing.

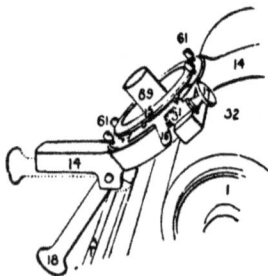

THE PHONOGRAPH CYLINDER.

UNTIL Mr. Edison invented the present What it is. metallic soap blank, which we will here- after call the wax cylinder because that is the common term applied to it, the mediums for recording were unsatisfactory. While the ingredients that compose this cylinder are not unknown in the art, the secret of their chemical combination still remains with Mr. Edison. Other blanks made in imitation of his and cast in the same form, lack the qualities essential for record- making, and other characteristics that give a cylinder permanent value. By the term blank, we refer to the blank cylinder; this cylinder, when it has been prepared with speech or music on it, is called a record, and these simpler terms will be frequently used.

As delivered by the factory, blanks do not have New blanks never shaved. surfaces prepared for record-making. They first require to be shaved to secure a smooth surface, no less than to true them in case they are eccentric. For this purpose a turning-off device is provided with the Phonograph. When once trued the blanks always remain cylindrical, but the surfaces must be re-shaved as often as new records are wanted. The operation of shaving cylinders and preparing sur- faces is fully explained under a separate heading.

The cylinder (32) is slipped on the Phonograph

How to put
on cylinder.

mandrel (1), thin end foremost, where it is held by friction, the inside of the cylinder being conical. A minute's practice is better than a page of instruction as to the proper pressure to cause the cylinder to bind. A warm cylinder should not be placed on a cold mandrel, as the sudden contraction of the wax may cause it to break, or if it does not break it will at least cause it to bind. By the same token, if a cold cylinder is placed on a warm mandrel, the cylinder will keep slipping, for heat from the warm mandrel communicated to the cylinder will expand and thus loosen it. Conditions like these do not arise if the cylinders and the machine are kept in the same or about the same temperature and room. The information is given for the benefit of slot-machine operators, and those persons who thoughtlessly bring in records on a cold day and put them on a machine which has been standing in a normal house temperature. Records left on a machine over night in cold weather are likely to be found tightly stuck or broken, from the same law of unequal expansion. If the cylinder is not broken, and is a blank or a valueless record, it can be easily freed by grasping it gently with the warm hand, and holding it until the heat of the hand causes the wax to expand. If the record is valuable, the simplest way to remove it is to place the entire Phonograph in a warm room. For slot Phonographs, more often exposed to such changes, a rubber mandrel is provided.

Co-efficient of
expansion.

To remove
a tight cylinder.

Careful
handling.

The Phonograph cylinder is fragile and brittle, and care must be exercised in handling. Cylinders are kept on pegs in boxes or trays to prevent them falling over, and to keep them from coming in contact with each other, as the slightest touch will injure the outside surface of a record. They are easily handled by thrusting the first and second fingers of the right hand into the thick end of the cylinder, and spreading the fingers apart. Never take hold of a record by

How to handle the cylinder.

the outside, or permit it to lie on its side. Although touching the surface of the record will not destroy it, the moisture from the hand will leave a mark on the surface that will be noticed in the reproduction, and any grit or fine dust will be pressed in, and cause a harsh, scratching reproduction. Dust and dirt are the foes of the record, and to keep cylinders as free from these as possible the camel's-hair chip brush is provided. With care records last a very long time, and though becoming somewhat harsh from constant wear, still can be reproduced many hundred times without appreciable difference. In packing records for transportation they are rolled in the finest quality of cotton batting, first splitting the cotton so as to bring the soft fibre next the cylinder. The glazed

side of the cotton must not be used next to the
surface. It will ruin the record. Oiled paper is now
wrapped around the cotton, and the cylinder placed
in a heavy pasteboard box made for receiving it
without crowding. These individual boxes are then
placed close together, and packed on an excelsior
cushion, with excelsior around and over them, and
boxed in the ordinary manner. The excelsior packing must not be too tight. Blanks having no
prepared surfaces are wrapped in tissue paper only,

Showing method
of wrapping.

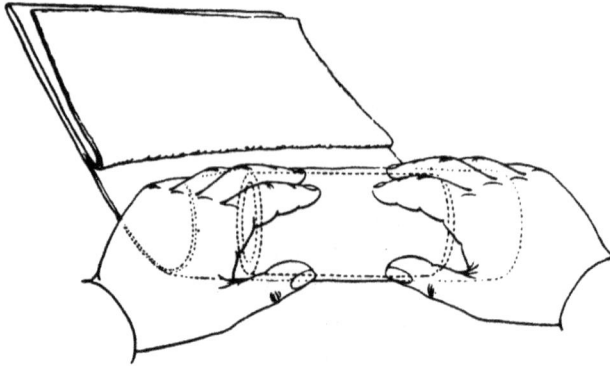

to prevent scratching, then rolled in excelsior.
Blanks are packed and sold in barrels containing 150
cylinders each, and are best shipped in that way.
Records can be more safely transported in barrels,
and large orders are invariably so shipped.

Color of blanks.   Blanks vary in shade, but the difference in color,
contrary to the popular impression, has nothing to do
with their wearing or recording qualities. The first
blanks Mr. Edison made were almost white, and later

they were very dark, while the medium and light chocolate shades are now generally supplied. The several colors are produced by different conditions of manufacture, and do not relate to the use.

Broken cylinders cannot be repaired. They have no commercial value. Broken cylinders of no value.

Cylinders, whether records or blanks, that go on one machine will fit any other machine. Cylinders interchangeable.

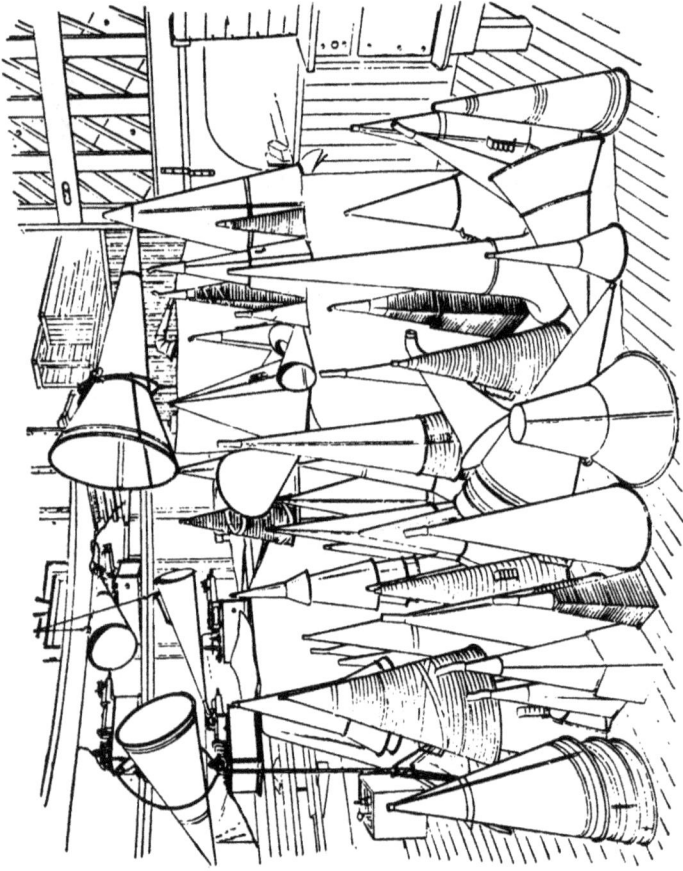

HORNS AND TUBES.

IN this chapter we do not expect to say all there is to say about horns, or to say the last word about horns, for the last word has not yet been spoken. The horn is still in its experimental stage, although certain definite results have been accomplished, and certain facts are known. One of the chief obstacles in the way is the fact now well proven that other conditions beside material and shape affect the practical action of the horn, especially in recording.

Horns still an experiment.

With the Phonograph a speaking tube and listening tube are provided. The speaking tube for dictation purposes meets the conditions acceptably. The single tube for listening is the best device for the purpose. But for concert use and public entertainment, the sound must be thrown out so that many persons can hear it, and for this purpose numerous types of amplifying horns have been produced. It would astonish the casual reader to learn of the number and thoroughness of the experiments in that direction. Mr. Edison has himself tried a vast number of sizes and shapes, out of all sorts of material. Other experimentalists and enthusiasts have gone over the same ground, and branched out into new paths. Yet all have come back to the main-travelled road. Wood, iron, steel, zinc, copper, brass, tin, aluminum, cornet metal, german silver, have been

Horn necessary.

Various materials employed.

tried. Glass, too, and hard rubber, papier-maché, an :
probably every other product that nature yields or
man contrives. The latitude as to form and shape
being greater than the resource in material, there
have been almost innumerable attempts in that line.
After all of which it may be said that tin and brass,
defective as they are, have been settled upon as the
most available, and the forms now known in the
trade as the most desirable. Any horn to be good
must come out of sound metal, and be perfectly joined.
Ordinary joining will not do, and imperfect metal is
a delusion.

Approved forms of horn. The horns in the trade have names which indicate
pretty clearly the character of each. First, there is
the small 14-inch tin horn, whose most valuable use
is for dictation and the making of talking records.
It has a flaring end opening away from the line of
the stem, is light, and easily slipped over the tube
plate of the speaker.   For reproduction it is one of the
most unsatisfactory horns made, yet a large number
of people insist upon using it for that service. Records
reproduced through it are likely to squeak and rattle,
and if, because of greater handiness, a small horn is
desired, one of the small nickel-plated brass ones
should be obtained. The test with the 14-inch tin
Standard tin horn. horn is the most difficult in the record business. The
26-inch standard tin horn is deservedly the amplifying
device most used, and all things considered, gives as
good results as any. It is not expensive, can be used
for recording and reproducing both, and fulfills all
reasonable requirements of horn service. When cor-
rectly made, block tin is used, and the joints are so

fastened as to prevent rattle. If made of cheap material, Must be properly made.
it is the same abomination that all other cheap supplies
for the Phonograph are. The horn is heavily japanned,
not for looks merely. It is held in place on a folding
tripod, to the loop of which it should be attached
by string, ribbon, or other non-conducting material,
never by a metal hook or wire. The connection with
the speaker of the Phonograph is effected by a short Connection with Phonograph.
length of rubber tubing. In the use of this, as with all
other large horns, the best results are obtained many
feet away from the mouth of the horn, which is so
built as to project the volume of tone forward. The
measurement at the bell or opening of this horn is
12 inches, and the lines from the bell to the nipple
are straight. Similar in results, but different in
character, is the 22-inch brass horn, preferred by Standard brass horn.
some because it is thought to give a more ringing
effect to the reproduction of band and orchestra music,
and claimed by others to make all reproduction
brighter. This horn has a flaring bell, and is 12 inches
in width at its mouth. It is suspended the same as
the 26-inch horn to the loop of a folding stand, and
makes a striking appearance.

Larger horns of various shapes are in the market, Large horns.
and invariably in tin or brass. Horns four feet or
more in length had a vogue at one time, but their
practical value was never apparent to the writer, nor
their musical superiority audible to his ear. For
advertising purposes they possess merit, and possibly
in heavy exhibition work, where it cannot be denied
that what their enterprising owners—some of them
" Professors," and all of them personal friends of Mr.

Edison, with whom they have formerly worked, or been acquainted since boyhood—call " putting up a great front," is effectually accomplished by these cavernous funnels. There are some larger horns, however, which possess merit and are not to be derided—horns larger than the standard types, and

The Philadel-
phia horn.

quite expensive. Mr. Fox, of Philadelphia, has devised a long horn of spun brass that is really excellent, if expensive, and other experimentalists are at work in the same field, to all of whom we wish well.

Recording
horns.

Recording horns are often wound with adhesive tape to check vibration, and make the tones of bass instruments more natural, or to give a ring to the bass register of a piano. A horn will sometimes make a much better record than it will reproduce. The best band records within our knowledge were taken regu-

Sometimes
inferior for
reproducing.

larly with a taped zinc horn 22 inches long, and 10 inches at the bell, but this horn is inferior for reproducing. One of the most perfect vocal-recording horns we ever saw was totally unfit to be used for reproducing. The true method, therefore, would appear to be to use that kind of horn which gives the best average results both for recording and reproducing.

To enable several persons to hear the reproduction of the Phonograph at the same time, without the use of a horn, an attachment with a number of tubes has

Multiple rails.

been devised. This multiple attachment is formed of hollow metal rails with from 14 to 17 outlets, to which single hearing tubes are attached, the whole fastened to the edge of the machine. The end is connected to the speaker of the Phonograph by a flexible tube, and the sound then distributed to the various tubes. In some kinds of exhibition work this attachment, which is called the way-tubes, is necessary, more especially where a charge is made for listening, and at all entertainments where outside noises would be likely to disturb the hearers. The sound is loudest through the tubes nearest to the speaker, although a good reproduction can be heard at all openings. The advent of the new duplex speaker, however, has now reinforced the sound at the end tubes. There are several other sorts of multiple attachments, with fewer hearing tubes. These generally are fastened directly to the speaker of the Phonograph by means of a metal hand, over the fingers of which single hearing tubes are slipped. The 3-way and the 4-way are the most popular forms.

Advantages of way-tubes.

The quality of tubing used should be that specially prepared for the purpose. The ordinary soft tubing of commerce will not do. It absorbs sound, and the tone becomes "dead." A hard-wall tubing with large hole is made for the purpose of better carrying the sound. This tubing is vulcanized to the highest point that will still leave it flexible and tenacious. The increased cost is nominal when compared with the results obtained.

Quality of tubing.

The interesting picture facing this chapter shows a group of recording horns used in a record laboratory. It was drawn from a photograph.

SHAVING OF CYLINDERS.

THE mechanical operation of the turning rest, and the way to shave cylinders, are explained in the chapter entitled Operation of the Machine. It is not proposed now to repeat those instructions, but rather to deal with shaving for special results.

Chip chute removed.

If much shaving is to be done, we advise that the chip chute (88) be removed and left off. Avoid touching the edge of the sapphire knife. Metal will fracture and destroy the cutting face. The advantage of having the chute off is that it does away with the clogging of the slot, removes the danger of injuring the sapphire while cleaning out the chute, prevents driving the wax particles back on to the surface when clogging begins, and allows freedom for adjustment of the knife,

Increased speed by throwing off belt.

as explained below. Ordinary shaving is done at the normal speed of the machine, which can be increased if desired by throwing off the governor belt, and letting the motor run wild, using the switch to start and stop the machine as before. The disadvantage is the untidiness that the loose wax chips or shavings cause. Professional shaving is done at a very

High-speed work.

high speed, from independent motive power, the best results coming from a maximum of 1500 revolutions per minute, while commercial high-speed shaving is carried on at about 2800 revolutions per minute. Above this limit, of say 3000 revolutions, it is not practicable to go, as sapphires cannot be ground to withstand the work, the vibration becomes excessive, and centrifugal force will disrupt the cylinder. *Maximum shaving speed.*

The cutting knife of the turning rest is of hardest sapphire, ground with diamond dust to a perfect edge —so delicate, in fact, that the slightest blow is likely to chip it. Hence care must be observed not to touch it with metal, and hence the danger from carelessly handling the chip trough. The shaving knife, shellaced in a brass holder, is affixed to the bar by a single screw. In ordinary work, unless the knife becomes loose, the adjustment will remain constant a long time. In severe work, the ordinary adjustment of this knife will not do, and it must be reset so that the cutting will be done on the left-hand side of the blade, the shaving rising from the knife at the left-hand corner. If the corner of the knife is set at too great an angle in its relation to the cylinder, it will cut and leave a heavy line. This spoils the value of the recording surface, by making it rough. We advise experiments in this matter, as no two knives work alike at high speed. The width of the shaving to be taken depends wholly upon the way in which the knife is ground, and has no direct bearing on the character of the surface, that can be forecast. No knife running at high speed will safely take as deep a cut as at lower speed, and deep cuts should not be attempted, for they break the surface. *Sapphire knife described.*

How to set knife.

Width of shaving.

A fruitful cause of bad surfaces is trying to shave with a chipped sapphire, the imperfection of which is not visible to the naked eye. A strong microscope will disclose the defect at once. Another one is lack *Chipped sapphires.*

of patience in adjusting the knife, combined with the probability, if on high-speed work, that the knife was not intended for that use. The ordinary knife is not suitable for such service, though it will do good work at the ordinary speed for which it was made. For holding the knife in place while screwing it into position, an open wrench that grips the sides of the brass holder close to the back, away from the sapphire setting, and clear of the cutting edge of the sapphire, can be easily made. In tightening the screw, after experiment has demonstrated the best position, care must be taken while tightening the set screw not to disturb the adjustment.

Wrench for adjusting knife.

Blanks vary in hardness according to length of seasoning, and a knife that will act well on a fresh cylinder will not always shave an old one properly. Always adjust on a blank of commercial thickness, or your experiment may prove a failure on thin blanks. The customary way of judging a shaved blank by looking at its surface is not the true way. The sight test is unreliable, and the sound test the only safe one. Shiny cylinders are not always the smoothest. They may appear to have a polish like glass, yet have a bad surface. We know of nothing more misleading than this superficial way of judging. When the blank has been shaved it should be put upon a machine before the next experiment, and the surface listened to through a short hearing tube at various places, when crackling or other defects, if any, will be discovered. The reproducing ball should be tracked back and forth within the limit of the speaker-adjust-

Varying hardness of blanks.

Sound not sight test best.

ing screw. This test ought to be maintained at intervals during the progress of shaving.

No expert who values his sapphires will use them *Gritty blanks.* on cheap and gritty blanks. Bear in mind that the sapphire is ground to a more perfect edge than a razor, takes a higher polish than steel, and is much more brittle. No sapphire however perfect will hold its edge and polish on such blanks, and inferior blanks never have a perfect recording surface, and never can have, and records made on them show this defect.

For all shaving, the body of a standard Phonograph *Machine for* is better than any independent mechanism. It may be *shaving.* connected to a small electric motor by proper belting, or driven from independent shafting. Care should be taken that none of the vibration from the actuating motor is imparted to the shaving device, or it will breed a crop of troubles too manifold to be recorded in a book of this size. The speaker should be removed from the speaker arm, and the speaker arm weighted so that it will ride steadily. Centers, shaft, *Oversight and* back rod, and nuts, must be liberally oiled, and the *renewal.* same oversight given to the perfect adjustment of every working part that is required of all other high-speed machinery. Owing to the velocity, the wear is enormous, and renewals must be made frequently.

Comprehensive as we have tried to make the foregoing, it covers, after all, only a small part of the field in this special department. Beyond and above what is written there remains unspoken the most vital word of all—WORK. Work, patient, untiring, intelli-

Work and
patience.

gent, ever enthusiastic—discouraging sometimes, it is true—will point the way, and we should not have felt that the chapter were complete without this disclosure.

SPEAKERS AND GLASSES.

THE speakers for the Phonograph are three *Speakers* in number: the standard, which records *described.* and reproduces, and the only one that records; the automatic, for reproducing only; and the duplex, also for reproducing. The standard speaker is a part of every complete machine, while the automatic and the duplex are sold as extras.

The management of the standard speaker in record- *Standard* ing and reproducing is explained in the general *speaker.* directions on another page. The reproducing stylus is adjusted to the track of the record as therein mentioned. Owing to the unequal expansion of the wax in different temperatures, and the rigid character of the weight required for recording, this adjustment is unavoidable and sometimes troublesome. To overcome this, the automatic speaker was invented by Mr. Edison. This speaker is now considered indis- *Automatic* pensable. It does away entirely with the use of the *reproducer.* adjusting screw, save occasionally to start the reproducing sapphire in its track. The self-adjustment overcomes the effect of the unequal expansion and contraction of the cylinder, and since the stylus always remains at the bottom of the record groove, the reproduction is louder and more perfect. On slot machines it is the only speaker to use, because it is the only one that insures a reproduction of the record

Advantages of
the automatic.

at all times; for exhibitors, because besides affording better results, it obviates the necessity of readjusting while a record is playing, and guards against the fading-away of the sound, so common to the standard speaker under unfavorable conditions; for private use, such as home enjoyment or transcribing, where reproduction is the chief office of the machine, it saves labor, simplifies handling, and yields more pleasing results.

The duplex speaker is an automatic reproducer, the diaphragm of which is enclosed in a chamber to

Duplex speaker.

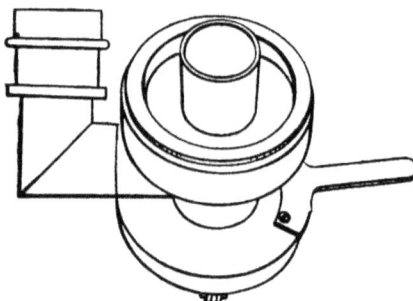

prevent the escape of the sound. It utilizes the vibrations from both sides of the glass, while in the other types of speaker the sound from the lower side of the glass escapes. When reproduction is heard from the upper opening, with lower or side opening closed, there is no perceptible outside sound. When the side opening is used in connection with the upper one, either with two horns or attached to multiple rails, the volume of sound is increased, and the carrying power strengthened.

Speakers require some attention, but not a great Care of
deal. The sapphires should be occasionally touched sapphires.
with benzine on the finger tip, to remove wax scales
and dirt, and in handling care must be taken not to
hit the sapphires with metal or any other hard sub-

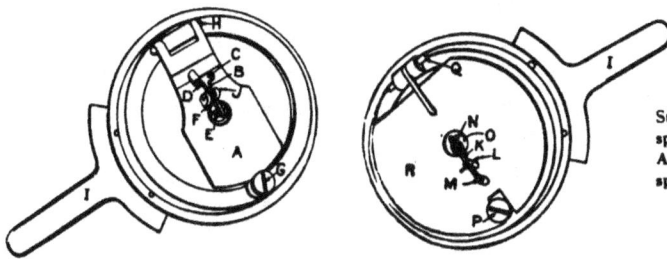

Standard
speaker (left).
Automatic
speaker (right).

		Index of speaker
A . . . Speaker Weight.	I . . . T-lever, or Cup Lever.	parts.
B . . . Recording-and-reproducing Arm.	K . . . Automatic Reproducing Arm.	
C . . . Sapphire Reproducing Ball.	L . . . Automatic Reproducing-arm Pin.	
D . . . Sapphire Recording Stylus.	M . . . Sapphire Reproducing Ball.	
E . . . Link.	N . . . Link.	
F . . . Crosshead.	O . . . Crosshead.	
G . . . Standard Limiting Screw.	P . . . Automatic Limiting Screw.	
H . . . Speaker-weight Hinge Pin.	Q . . . Automatic-weight Hinge Pin.	
I . . . T-lever, or Cup Lever.	R . . . Automatic speaker Weight.	
J . . . Standard Reproducing-arm Pin.		

stance. When speakers are removed they should be
laid face upwards, or put in a box with a place prepared
for holding them. If wax shavings from recording
accumulate on the sapphire bar, remove them to
prevent gumming. The diaphragm is held in place Renewal of
by two rubber cushions called the gaskets. These gaskets.
cushions need renewal when they become hard or
gummy.

The diaphragm, which is really the talking
part of the speaker, was adopted after experiment
with every available material. Originally of light

sheet tin, it was followed by mica, then shellaced bolting silk, and finally glass; while experiments have been made with ivory, celluloid, hard rubber, skins, paper (there is one concern we lately heard of that is putting out a disc cut from an ordinary business card, under the pretentious name of a " fibre diaphragm,") aluminum, ground steel, silver foil, etc. The character of the glass used has all to do with the

Different kinds of glasses.

clearness and brilliancy of reproduction. Glasses are produced in several grades. The cheaper grades seem to lack firmness, and are unresponsive, and the reproduction they afford is deficient in musical color. Parts of the same diaphragm will

Best glasses only to be used.

vary in thickness and be untrue. The best selected glasses of foreign manufacture are the only ones to use, and the slight additional cost is an insignificant item. They are made carefully, tempered to the utmost flexibility, and prove uniform in quality. Yet the selection even of these is difficult. We recently saw 30 accepted out of 140 submitted, the balance being destroyed.

Thickness of glasses.

The standard speaker requires a slightly heavier glass than the automatic or the duplex. No fixed rule can be laid down, as glasses must be chosen with a view to the service contemplated.

How to set glass in speaker.

The glass is set in the speaker in the following manner. The speaker must be taken out of its seat in the speaker arm by removing the clamps (38) which hold it. Unscrew the collar which holds down the tube plate (89) on top of the speaker, and lift out the tube plate. Then remove broken glass and rubber gaskets. See that no old rubber is left to adhere to the edge of the metal. Set new gaskets in place with the glass between, replace tube plate and collar. Unscrew the limiting screw (G), and raise the floating weight, taking care not to lose the link (E) and the crosshead (F). Place a small drop of stratena, which is a strong fish glue, in the center of the glass. This can be applied on the end of a toothpick or point of a pin. Then bring the weight back to place, setting the limiting screw as the weight comes down so as to bring the lower collar of the screw under the projecting end of the weight. Set the crosshead in the stratena in the center of the glass, taking care that the link rests in its natural position without binding; gently press the crosshead on the glass, and hold it there by forcing a small wedge, such as the tapered end of a match, under the sapphire end of the recording and reproducing bar. This will throw the pressure of the weight on the crosshead, where it should remain until dry. Natural drying, which is the best, requires several hours. A quicker but more dangerous way of drying the stratena is to hold the head of a heated nail against the opposite side of the glass, under the crosshead, on which press down with the finger while pressing up against the glass on the other side. The danger of this operation is that the glass is likely to crack on the sudden application of heat, and the stratena, if burned, will lose its adhesiveness.

To affix the crosshead.

RECORD-MAKING.

State of the art.

THE making of records, in the sense of recording sound on a prepared blank cylinder, and the making of records for commercial purposes, are two distinct processes. The operator with scarcely any special knowledge can prepare a distinct and satisfactory record by following the general directions which accompany every instrument. But musical and other records intended for sale to the public require to be made to meet the conditions which the market exacts. A familiar illustration would be the difference between amateur and professional work on the stage, or in one of the crafts. Proficiency in professional record-making has been acquired after years of application, by men who devote their whole time to the art, and who are still learning.

Standard speaker used.

The regular standard speaker is the speaker used for record-making, and the results attained, while they may seem different from the ordinary records that users themselves produce, are reached by adapting the speaker to the particular use for which it is desired, and making these changes as frequently as circumstances require. This is what

Building-up a speaker.

is known as building-up the speaker, although in practice a large number of speakers are used, so that when good results are once obtained under certain conditions with certain speakers, the same

speakers are seldom if ever used for any other purpose. Speakers are built up to meet different varying conditions, such as voice, tone, atmosphere,
acoustic surroundings, etc.

Since the principal functions of the speaker are
performed by the recording sapphire and the glass
diaphragm, these parts require the most careful
attention. The sapphire should present a perfectly
sharp edge to the cylinder, and this edge must be
maintained under all circumstances. As stated elsewhere, the surface of the blank should be smooth, or
it will not be possible to cut a clean record in the
wax, no matter how perfect the sapphire. If the
cylinder is smooth and the record rough, examination
under a glass will probably discover a chipped edge
on the sapphire. The remedy is to turn the sapphire
in the socket, which is done by heating the socket
until the shellac holding the jewel is softened. The
sapphire must be replaced at the same angle after
turning it, and the turning should be done without
withdrawing the jewel from its setting. Tweezers
with the ends hollowed out to hold the point, afford a
convenient means of making this adjustment. A dull
sapphire has no value in record-making, owing to the
hardness of the medium, and the slight power trans-
mitted from the glass. A dull point not only will not
cut as well, but it offers resistance to the work
required of the cutting edge. Realize how delicate
are some of the indentations to be cut, and the requirements can be understood. Experiment with the angle
of the sapphire will be found to produce changes in
the clearness of the record, and this should be studied

Recording sapphire.

Turning the recording stylus.

Dull sapphires useless.

Action of the
weight.

as the changes occur. A floating weight (*A*) holds the sapphire to proper depth. Nothing is gained by increasing this weight, as was formerly supposed, while the loss of tone is considerable, owing to the restriction on the free action of the cutting point, with a resultant harsher surface. The weight must hang freely on its hinge, and the sapphire recording-and-reproducing arm freely on its fulcrum.

Selecting glasses
for record work.

The selection and adjustment of glasses present a complex problem. Only the clearest and most brilliant are chosen for recording. Soft, spongy glasses, without resilience, unfitted as they are for adequate reproduction, are practically valueless for recording. It is necessary to get glasses that will respond to the peculiarities of a voice, or the qualities of an instrument, to make a faithful record of that voice or instrument.

Making vocal
records.

With regard to vocal records, it may be said that different singers rarely present the same conditions to the record-maker, and more than that, the same singer presents different conditions during different stages of his work. To meet these physical changes, new diaphragms are required, or a different position exacted with respect to the horns, if the highest results are expected. Speaking generally, the singer is placed about a foot-and-a-half from the central horn, and the piano, if an upright, at right-angle to the line of the horns, and preferably raised so that the horns will center in the middle of the sounding-board facing the machines. Never use the loud pedal of the piano.

Instrumental
records.

In the case of orchestra and band work, the performers should be arranged so that those instruments which carry the melody are not subordinated to the secondary parts, yet with the aim of bringing out every instrument at its proper musical value.

Heavier glasses are used for band and orchestra work, and for loud solo instruments, while thinner glasses, gauged to the condition of the singer's voice, are invariably required in vocal work. No fixed rule can be laid down, and experiments must be conducted

Different glasses required.

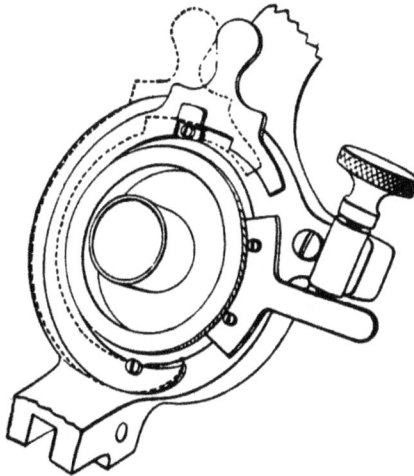

on the above general lines until satisfactory results are reached. There is no royal road to record-making, and those who expect to make commercial records by rule of thumb will be disappointed, unless they succeed in discovering one of those secrets which so far have eluded the grasp of the best experimentalists. The art of record-making has not yet developed laws which may be set down as sure guides.

Rapid clamping device for changing speakers.

Rules as to
thickness of
glass.
In adjusting speakers for special work, it will be found that experiments indicate several things positively, and we will give this information as clearly as may be. A heavy baritone voice is best recorded by a glass that measures about 6½ thousandths of an inch in thickness, in combination with a rubber horn connection about an inch-and-a-half in length. This is a very sensitive glass, and is necessary in order to record faithfully the weaker tones of the lower register. The average tenor voice needs a thinner glass, and on the open vowels the singer must be trained to withdraw from the mouth of the horns to avoid blasting. Heavier glasses are used for bands, measuring from 8 to 9 thousandths of an inch. Even thicker diaphragms than these give excellent results, for absolute firmness is essential to prevent over-vibration.

Changing horns
and connections.
These rules as to thickness of glass indicate the general practice, which is most likely to be followed with success. Assuming that the glass is right, variation in the size of the receiving horn will often bring the record to the desired strength and quality. In general, the volume and tone is thinned by decreasing the diameter of the horn, and made oral by increasing the width at the bell. A variation in the length of the rubber horn connection will soften the tone, and sometimes accomplish what is sought when a change of glass will not, provided a glass has been found which gives perfect mechanical results but is deficient in musical quality. Economy cannot be practiced in the use of glasses when efficiency is desired.

A number of records are generally taken at the

same time, care being observed to focus the singer or performers at the common center of the horns.

<small>Several records made at once.</small>

So much has been said about the so-called "cheap" records that it is not worth while to go into that matter at any length. The cost of a record, and therefore its value, depends upon the labor employed in its making, and the more the labor and the greater the intelligence the more a record is worth, because the more accurate the result. These conditions preclude the making of good records at a low price in the present state of the art. A record to be good must be true to nature, and possess tone and musical color, both of which features are absolutely lacking in low-grade product. The common defect of such records is a wiriness of tone, combined with a loss of that sympathetic quality which gives music its principal charm. In view of the fact that the Phonograph will yield results of a very high musical order, the parrot-like reproduction of rhythm and form merely, cannot be seriously considered in discussing the question of records.

<small>Cost and value of a musical record.</small>

<small>Cheap records a delusion.</small>

There are many other topics in regard to the uses of the Phonograph that might be properly discussed in a more ambitious treatise on the subject. In the limited space which the plan of a work of this kind imposes, it is possible only to touch, and that briefly, on the more obvious features of the Phonograph and its work which come within the wants of an every-day experience. Researches into the scientific features of Mr. Edison's invention, the properties of

<small>L'envoi.</small>

sound, the phenomena that arise in connection with their application in recording and reproducing, are matters for the trained scientist, and beyond the scope of a practical guide.